T0206085

Data-Driven Farming

In the dynamic realm of agriculture, artificial intelligence (AI) and machine learning (ML) emerge as catalysts for unprecedented transformation and growth. The emergence of big data, Internet of Things (IoT) sensors, and advanced analytics has opened up new possibilities for farmers to collect and analyze data in real-time, make informed decisions, and increase efficiency. AI and ML are key enablers of data-driven farming, allowing farmers to use algorithms and predictive models to gain insights into crop health, soil quality, weather patterns, and more. Agriculture is an industry that is deeply rooted in tradition, but the landscape is rapidly changing with the emergence of new technologies.

Data-Driven Farming: Harnessing the Power of AI and Machine Learning in Agriculture is a comprehensive guide that explores how the latest advances in technology can help farmers make better decisions and maximize yields. It offers a detailed overview of the intersection of data, AI, and ML in agriculture and offers real-world examples and case studies that demonstrate how these tools can help farmers improve efficiency, reduce waste, and increase profitability. Exploring how AI and ML can be used to achieve sustainable and profitable farming practices, the book provides an introduction to the basics of data-driven farming, including an overview of the key concepts, tools, and technologies. It also discusses the challenges and opportunities facing farmers in today's data-driven landscape. Covering such topics as crop monitoring, weather forecasting, pest management, and soil health management, the book focuses on analyzing data, predicting outcomes, and optimizing decision-making in a range of agricultural contexts.

Data-Driven Farming
Harnessing the Power of AI and Machine Learning in Agriculture

Edited by
Dr. Syed Nisar Hussain Bukhari

CRC Press
Taylor & Francis Group
Boca Raton London New York

CRC Press is an imprint of the
Taylor & Francis Group, an **Informa** business

AN AUERBACH BOOK

First edition published 2024

2385 NW Executive Center Drive, Suite 320, Boca Raton FL 33431

and by CRC Press

4 Park Square, Milton Park, Abingdon, Oxon, OX14 4RN

CRC Press is an imprint of Taylor & Francis Group, LLC

ISBN: 978-1-032-61892-0 (hbk)
ISBN: 978-1-032-77872-3 (pbk)
ISBN: 978-1-003-48517-9 (ebk)

DOI: 10.1201/9781003485179

Typeset in Minion
by SPi Technologies India Pvt Ltd (Straive)

This book is dedicated to my loving father, the late Syed Gulam Mohi ud Din Bukhari (Aba Jan) and my loving and caring mother, the late Zareefa Syedah, whose wisdom and love continue to inspire my quest for knowledge and innovation.

Contents

Editor Biography

Dr. Syed Nisar Hussain Bukhari is an accomplished professional with a strong background in academia, research, and administration. He is currently serving as Scientist-D at NIELIT Srinagar and has made significant contributions to the field of information technology (IT).

Dr. Bukhari obtained his bachelor's and master's degrees in computer applications from the University of Kashmir, Srinagar, India, and PhD in the area of machine learning, bioinformatics, and immunoinformatics from the University Institute of Computing, Chandigarh University, India.

Dr. Bukhari's research interests encompass various cutting-edge fields, including machine learning and deep learning, bioinformatics, and immunoinformatics. His work has been recognized in prestigious SCI journals such as *Nature Scientific Reports*, *Springer*, *MDPI*, and *Hindawi*, reflecting the significance and impact of his contributions to the scientific community.

Dr. Bukhari's commitment to imparting knowledge is evident through his extensive teaching experience. He has taught various core subjects, including machine learning, Python, data structures, web programming using ASP.Net, and more to postgraduates – namely, MCA and MSc IT. His comprehensive knowledge and skills have allowed him to impart training and internships in artificial intelligence, machine learning, deep learning, and other emerging technologies to engineering students.

With a passion for knowledge dissemination, Dr. Bukhari has published numerous papers in reputed journals and conferences. Dr. Bukhari's academic achievements extend beyond research publications. He has edited and authored books and received recognition as a certified machine learning expert from Stanford University and Johns Hopkins University. Additionally, he has been invited as a speaker to various seminars, webinars, and conferences, showcasing his expertise in the field. His commitment to excellence has been acknowledged with the receipt of the Best Paper Awards at multiple international conferences.

His expertise has been sought by prestigious journals like *Briefings in Bioinformatics* by Oxford University Press, *Computers in Biology and*

Medicine by Elsevier, and *Scientific Reports* by Nature, where he serves as an invited reviewer.

As a session chair, he has chaired various special sessions in Scopus-indexed international conferences. His research interests include artificial intelligence and machine learning, deep learning, bioinformatics, immunoinformatics, computational biology, and web security.

In his role as the Head of the Department of IT at NIELIT Srinagar, Dr. Bukhari demonstrated exceptional leadership and played a vital role in the growth and development of the department. During his tenure, he played a pivotal role in the department's growth and development, overseeing the MSc IT and BSc IT courses. He has provided consultancy services to different departments of the Government of Jammu and Kasmir, India, offering his expertise in implementing various e-governance-related projects.

Dr. Bukhari holds memberships with the Institution of Electronics and Telecommunication Engineers (IETE) and the International Association of Engineers (IAENG).

Contributors

Nelofar Ara
Lovely Professional University
Phagwara, India

VK Aswathy
Lovely Professional University
Phagwara, India

Saimul Bashir
Chandigarh University
Mohali, India

Mohammad Ubaidullah Bokhari
Aligarh Muslim University
Aligarh, India

Syed Nisar Hussain Bukhari
National Institute of Electronics
and Information Technology
(NIELIT)
Srinagar, India

Muneer Ahmad Dar
National Institute of Electronics
and Information Technology
(NIELIT)
Srinagar, India

Deepthi Das
CHRIST (Deemed to be
University)
Bangalore, India

Faisal Firdous
Jaypee University of Information
and Technology
Solan, India

Anshika Gupta
Dr. Ambedkar Institute of
Technology for Handicapped
Kanpur, India

Amarthya Dutta Gupta
CHRIST (Deemed to be
University), Kengeri Campus
Bangalore, India

Uzma Hameed
Govt. College for Women,
Cluster University
Srinagar, India

Inzimam Ul Hassan
Vivekananda Global University
Jaipur, India

Ummer Iqbal
National Institute of Electronics
and Information Technology
(NIELIT)
Srinagar, India

Kevin Johnson
CHRIST (Deemed to be
University), Kengeri Campus
Bangalore, India

Kalpana Katiyar
Dr. Ambedkar Institute of
Technology for Handicapped
Kanpur, India

Devinder Kaur
Sri Guru Granth Sahib World
    University, Mata Gujri College
Fatehgarh Sahib, India

Muhammad Najeeb Khan
Shri Mata Vaishno Devi University
Katra, India

Aakansha Khanna
Chandigarh University
Ludhiana, India

Gursimran Jeet Kour
Chandigarh University
Mohali, India

S. Babu Kumar
CHRIST (Deemed to be
    University), Kengeri Campus
Bangalore, India

Kukatlapalli Pradeep Kumar
CHRIST (Deemed to be
    University), Kengeri Campus
Bangalore, India

Arya Kumari
Shri Mata Vaishno Devi University
Katra, India

Sani Mustapha Kura
Lovely Professional University
Phagwara, India

V.S.S.P.L.N. Balaji Lanka
Vignan Institute of Technology and
    Science
Hyderabad, India

P. Laxmikanth
Vignan Institute of Technology and
    Science
Hyderabad, India

Manoj Kumar Mahto
Vignan Institute of Technology and
    Science
Hyderabad, India

Rukaya Manzoor
National Institute of Electronics
    and Information Technology
    (NIELIT)
Srinagar, India

Hamira Mehraj
Govt. College for Women, Cluster
    University
Srinagar, India

Meghan Mary Michael
CHRIST (Deemed to be
    University), Kengeri Campus
Bangalore, India

Thamizhiniyan Natarajan
Gandhigram Rural Institute
Chinnalapatti, India

A. Nidin
CHRIST (Deemed to be
    University)
Bangalore, India

Sana Farooq Pandit
National Institute of Electronics
    and Information Technology
    (NIELIT)
Srinagar, India

Shanmugavadivu Pichai
Gandhigram Rural Institute
Chinnalapatti, India
Srinagar, India

Qurat-ul-ain
Govt. College for Women, Cluster
    University
Srinagar, India

Amit Kumar Sinha
Shri Mata Vaishno Devi University
Katra, India

Mohit Soni
Dr. Ambedkar Institute of
    Technology for Handicapped
Kanpur, India

A. Vijayalakshmi
CHRIST (Deemed to be University)
Bangalore, India

Amandeep Kaur Virk
Sri Guru Granth Sahib World
    University
Fatehgarh Sahib, India

Mohammad Amin Wani
Lovely Professional University
Phagwara, India

Jewiara Khursheed Wani
National Institute of Electronics
    and Information Technology
    (NIELIT)
Srinagar, India

Gaurav Yadav
Aligarh Muslim University
Aligarh, India

Md. Zeyauddin
Aligarh Muslim University
Aligarh, India

Preface

It is with great pleasure and enthusiasm that I present to you *Data-Driven Farming: Harnessing the Power of AI and Machine Learning in Agriculture*. This book stands as a testament to the dynamic intersection of technology and agriculture, an evolution that has significantly transformed the way we approach farming practices. In recent years, the integration of artificial intelligence (AI) and machine learning (ML) into the agricultural landscape has marked a new era—a progressive shift from traditional methods to more sophisticated, efficient, and sustainable approaches.

The genesis of this book stems from a growing recognition of the pivotal role technology plays in addressing the complex challenges faced by the agricultural industry. The need for increased food production to meet the demands of a burgeoning global population, coupled with the imperative to manage resources more sustainably, has spurred a revolution. The utilization of AI and ML presents a compelling solution, enabling farmers, researchers, and stakeholders to optimize production, streamline processes, and make informed decisions crucial for the future of agriculture.

This book is a collaborative effort, encompassing the expertise and insights of various contributors from diverse fields—agricultural science, data science, technology, and beyond. Their collective wisdom and experiences have been meticulously crafted into a comprehensive guide that explores the multifaceted applications of AI and ML in agriculture. Readers will discover a wealth of knowledge, practical applications, and case studies that illustrate the transformative impact of data-driven approaches in farming.

The chapters contained within this book offer an in-depth exploration of the symbiotic relationship between AI, ML, and agriculture. From leveraging the Internet of Things (IoT) for precision health monitoring in livestock to the deployment of deep learning for crop disease detection, the depth and breadth of knowledge shared by our contributors demonstrate the versatility and promising implications of these technologies in agriculture.

This book is intended for a diverse readership—academics, researchers, farmers, technology enthusiasts, policymakers, and anyone interested in the transformative power of technology in agriculture. I hope that the insights presented within these pages will serve as a valuable resource, igniting curiosity, fostering innovation, and guiding the evolution of agriculture into a more sustainable and efficient future.

I extend my heartfelt gratitude to all the contributors whose dedication and expertise have made this book a reality. I also express my appreciation to CRC Press, Taylor & Francis Group, for their support and commitment to disseminating knowledge in this ever-evolving field.

Warm Regards,
Dr. Syed Nisar Hussain Bukhari
Editor

Acknowledgments

The compilation of *Data-Driven Farming: Harnessing the Power of AI and Machine Learning in Agriculture* has been a journey filled with collaboration, support, and dedication from a multitude of individuals and organizations. The completion of this book would not have been possible without the invaluable contributions and unwavering support from numerous places.

First and foremost, I would like to submit all praises and thanks to **Almighty Allah**, the most beneficent and merciful, who blessed me with strength, protection, and patience to overcome all odds and made this work to come to fruition.

I extend my sincere gratitude to all the contributors, whose expertise, dedication, and passion have shaped this book. Each chapter is a testament to their commitment to advancing the knowledge frontier in the dynamic field of data-driven agriculture. Their insights, research, and real-world experiences have been instrumental in crafting a comprehensive resource for the industry, academia, and all stakeholders interested in the intersection of technology and agriculture.

I would also like to express my deepest appreciation to CRC Press, Taylor & Francis Group, for their continuous support, guidance, and belief in this project. Their commitment to disseminating knowledge and fostering the publication of innovative works has been pivotal in bringing this book to fruition.

Additionally, I am profoundly grateful to the reviewers and editors who dedicated their time and expertise to ensuring the academic rigor and quality of this publication. Their insightful feedback and meticulous review have undoubtedly enhanced the overall standard of the content, enriching the readers' experience and the credibility of the book. The support and encouragement from academic and research institutions have played a crucial role in shaping this book. I express my appreciation to all the institutions that have fostered an environment conducive to research, innovation, and the exploration of cutting-edge technologies in agriculture.

I extend my heartfelt thanks to the farmers, agricultural experts, and industry professionals whose practical insights and experiences have been the backbone of this book. Their hands-on knowledge and willingness to adopt new technologies have inspired and guided the direction of this publication, ensuring its relevance and practicality in real-world agricultural settings.

I also express my gratitude to the readers of this book. I sincerely hope that the information and insights contained within these pages serve as a valuable resource, igniting curiosity, inspiring innovation, and fostering a deeper understanding of the powerful applications of AI and ML in agriculture.

Last but not least, I am deeply grateful for the unwavering support and inspiration provided by my late father and mother. Their enduring influence, values, and encouragement continue to resonate in every aspect of my life and work. Their belief in the pursuit of knowledge, their unwavering support during my endeavors, and their enduring love have been the guiding force behind this work. I carry their teachings of resilience, integrity, and the value of continuous learning as guiding principles that have shaped this project. Thank you to everyone who has contributed, supported, and been part of this endeavor. Your collective efforts have been instrumental in the creation of this comprehensive resource, and I am deeply appreciative of your involvement in this significant project.

Warm Regards,
Dr. Syed Nisar Hussain Bukhari
Editor

1

Leveraging IoT for Precision Health Monitoring in Livestock with Artificial Intelligence

Devinder Kaur

Sri Guru Granth Sahib World University, Mata Gujri College, Fatehgarh Sahib, India

Mata Gujri College, Fatehgarh Sahib, India

Amandeep Kaur Virk

Sri Guru Granth Sahib World University, Fatehgarh Sahib, India

1.1 INTRODUCTION

India, a densely populated nation, relies significantly on agriculture and dairy farming as a primary source of income, particularly within rural communities. The rapid population growth underscores the necessity for increased future food and dairy production. Historically, most Indian states have adhered to traditional livestock farming methods, with farmers making decisions based solely on their personal experience. However, contemporary trends are shifting away from these customary approaches as precision dairy technology gains prominence, propelling dairy farmers toward the realm of precision livestock farming. To make informed choices in livestock management, the availability of quantitative real-time data is crucial. This type of data is accessed and studied through the utilization of Information and Communication Technology (ICT), data analysis, machine learning (ML), and control systems [1].

Since humans rely on animals to meet their dairy product needs, it becomes crucial to uphold the well-being of these animals. Given that

DOI: 10.1201/9781003485179-1

certain diseases can be transmitted from animals to humans, the early prediction of cattle health and diseases holds significant importance [2]. However, the various facets associated with dairy farming, including milking, feeding, reproduction, and health, are frequently treated as distinct entities. As a result, ensuring a healthy life for each individual animal has become a formidable task. In the current era, innovation plays a pivotal role in healthcare systems. Among the domains experiencing significant advancements in recent years, electronic dairy farming stands out as particularly important [3]. Modern technologies, such as sensors, big data, and ML, present a fresh opportunity for farmers. Rather than reacting to diseases only when they become apparent, these technologies enable continuous monitoring of crucial animal well-being indicators, including movement, temperature, pulse rate, and humidity. Through the continuous data collection facilitated by sensors and the predictive capabilities of ML, farmers can now proactively identify, anticipate, and prevent cattle diseases prior to wide-scale outbreaks and the resulting catastrophic losses.

This kind of system offers a dual benefit to farmers. First, it enables a smaller number of farmers to efficiently manage a larger number of animals, leading to reduced production costs. Second, the system has the capability to notify farmers about potential diseases even before they reach the clinical stage. This early alert empowers farmers to take timely measures [4]. These innovative techniques applied in the dairy farming sector introduce novel approaches that can enhance and elevate future decision-making processes.

Despite the advancements in research, the practical application of these technologies has not been fully realized. Currently, veterinary doctors predominantly assess the physical parameters of animals through manual methods. Livestock farmers encounter significant challenges when it comes to monitoring cattle health. This underscores the critical necessity of translating theoretical knowledge into functional real-world systems.

1.2 CONCEPT OF PRECISION LIVESTOCK FARMING

Precision livestock farming (PLF) is a dynamic approach to monitoring that involves the continuous observation of animals through various methods, such as behavior monitoring, analysis of milk constituents and yield, video analysis, record assessment, and physiological monitoring [5]. PLF involves continuous health monitoring of animals using methods like camera and image analysis, sound analysis, and sensors. These tools can be wearable, integrated with milking machines, stand-alone, or part of management software, all aimed at enhancing animal health in precision dairy technology [6].

The inception of PLF can be traced back to the introduction of individual electronic milk meters for dairy cows during the 1970s and early 1980s. This was followed by the implementation of behavior-based estrus detection, rumination tags, and real-time online milk analyzers [7]. Many of the pioneering developments in PLF originated in Europe, notably in countries like the United Kingdom, Belgium, Germany, Denmark, the Netherlands, Finland, and Israel [1].

PLF presents a promising avenue for achieving sustainable livestock production, necessitating collaboration across diverse research disciplines. The objective is to translate technological advancements into practical, actionable insights for farmers, equipping them to make well-informed choices [8].

Currently, PLF predominantly benefits larger animals, such as dairy cows, beef cattle, and horses, where the economic returns justify the investment in monitoring tags [7]. Dependable PLF solutions can elevate animal welfare and productivity, driving advancements in the field to the advantage of both livestock and their overall well-being [9]. As PLF technology becomes more cost-effective, its potential to provide additional value for stakeholders, especially farmers, becomes evident. It facilitates enhancements in animal welfare, health, efficiency, and environmental impact [10]. By combining hardware and intelligent software, valuable insights are extracted from diverse data sources, empowering farmers to improve animal health, well-being, yields, and environmental contributions [11]. One of the most important goals of PLF is to work on the well-being of animals [12]). Therefore, the PLF technologies offer the chance to further develop efficiency and recognize medical problems at the beginning phase.

These "smart technologies" designed to assist livestock farmers in monitoring animal well-being are currently a prominent subject of research. Numerous countries are already allocating significant resources to their development, with the aim of transitioning toward more sustainable practices in livestock farming.

1.3 IMPORTANCE OF MACHINE LEARNING IN PRECISION LIVESTOCK FARMING

Within dairy research, ML emerges as a promising technique with the potential to aid farmers in improving decision-making. ML algorithms possess the capacity to analyze integrated datasets, thereby facilitating holistic livestock management that encompasses essential influencing factors within the complex system of cattle, including cows [13]. By employing ML algorithms, it becomes feasible to predict the standard behavior of animals and issue alerts when observed behavior deviates from predefined thresholds.

In the domain of cattle husbandry, ML has been effectively employed to predict outcomes such as reproductive results, high somatic cell counts, and the onset of calving. The utilization of "big data" in farm animal medicine has expanded, and its conversion into "smart data" is gaining traction, enabling the maximization of existing data resources. Additionally, professionals currently invest substantial time in the analysis and interpretation of data to arrive at clinical diagnoses. While a human may require 30–60 minutes to comprehensively analyze all available data and identify ailments like mastitis, an ML system can accomplish this task within seconds. This rapid, straightforward diagnostic procedure can be performed frequently, leading to more consistent diagnostic reports [14]. ML algorithms can store category data and analyze enormous datasets that are difficult to evaluate using traditional statistical methods. As a result, statisticians claim that ML algorithms produce better results because they learn from the data presented, whereas traditional analysis methods are skewed by the researcher's hypothesis [13].

Hence, ML holds significant potential as a valuable tool for PLF, aiding farmers in safeguarding the welfare of their animals.

1.4 LITERATURE REVIEW

In this section, we are going to present the literature review to explore the amalgamation of the Internet of Things (IoT) and ML in the domain of livestock farming. By analyzing a variety of scholarly works, this review aims to elucidate the current research landscape, advancements, and applications at this innovative intersection. Our objective is to emphasize the significance of IoT and ML in enhancing livestock management, health monitoring, and productivity. Through this examination, we pinpoint emerging trends, challenges, and prospects in this evolving field, offering insights into potential avenues for future research and development.

Swain et al. [15] introduced a real-time application device that enables farmers to monitor and assess the present health metrics of cattle by comparing them against established healthy parameters, serving as a baseline to identify potential declines in animal well-being. The system utilized various sensors, including humidity and temperature sensors, pulse sensors, and rumination sensors, along with the Xbee module and Arduino 1.0.6 software. The outcomes were successful, achieving an accuracy rate of 72%–75% in detecting the health status of cows.

In a study conducted by Borchers et al. [16], automated activity, lying, and rumination monitors were employed to assess prepartum behavior and forecast calving in dairy cattle. The research evaluated the utility of ML approaches such as linear discriminant analysis, Random Forest, and neural networks in predicting calving. Through the utilization of neural network techniques, a combination of activity, rumination time, and lying behavior variables was analyzed, leading to the generation of precise and sensitive alerts.

Liakos et al. [17] presented a novel integrated computational analysis for predicting lameness in cattle using ML techniques. The study incorporated factors such as steps per day, total walking per day, lying duration per day, and eating frequency per day as contributing elements. The newly developed algorithm was tested on datasets containing both healthy and unhealthy cattle samples. The primary objective of utilizing these four features was to effectively distinguish and separate the samples. Among the ML methods employed, Random Forest yielded the most promising results.

Cowton et al. [18] devised and assessed a deep-learning approach employing Particle Swarm Optimization (PSO) to monitor the health of cattle. This methodology was specifically employed by leveraging environmental sensor data to enable the early identification of respiratory illness in developing pigs. An autoencoder model was constructed using two recurrent neural networks (RNNs). The results of the study highlighted that alterations in the environment could induce signs of respiratory illness in pigs within a remarkably short span of 1–7 days.

Kumari and Yadav [19] employed a Raspberry Pi 3 model as a central controller to investigate crucial indicators affecting cattle health, including body temperature, heart rate, and rumination. Through the utilization of IoT, a cloud-based database was established, enabling the captured data to be accessible online from any location. This data could be accessed by a farmer or dairy owner using an Android application on their smartphone, or alternatively, they could receive an email notification for easy access to monitor their animals' health.

Pratama et al. [20] proposed the development of a collar device capable of capturing body temperature, heart rate, and movement of cattle. In addition, they designed a base station for local server management and a web application to display data, facilitating the analysis of cattle health issues. To classify health outputs into normal, slightly abnormal, and severely abnormal categories using data collected from the sensors, ML algorithms were employed.

Benjamin and Yik [21] elucidated various ML algorithms, provided a comprehensive overview of existing literature, and deliberated on the potential applications of these technologies for enhancing pig health. Their focus centered on essential sensors and sensor network systems that could contribute to advancements in this area. Vyas et al. [22] used the IoT to detect foot and mouth illness and mastitis in cows. ML techniques were used to detect these disorders using a range of sensors to distinguish numerous factors in animals, such as temperature, motion, and sound.

Xu et al. [23] aimed to investigate the feasibility of clustering the metabolic status of individual cows in early lactation based on plasma values. They also evaluated the effectiveness of ML algorithms in predicting metabolic status using on-farm cow data. The study analyzed eight commonly used ML algorithms, including K-Nearest Neighbor, Support Vector Machine, Decision Tree, Naive Bayes, Bayesian Network, Artificial Neural Network (ANN), Bootstrap Aggregation, and Random Forest. From their analysis, the two most successful algorithms in predicting metabolic status using on-farm data were the Support Vector Machine and Random Forest.

Chaudhry et al. [2] conducted a study that involved contrasting the features offered by various systems alongside their constraints. The research also encompassed an evaluation of existing technology-driven solutions and the associated equipment. Furthermore, the study introduced a real-time animal health monitoring system grounded in the IoT. Garcia et al. [24] conducted a systematic literature review that focused on the recent progress in utilizing ML within the context of PLF, with a specific emphasis on grazing practices and animal health. The review underscored the potential applications of ML in the livestock industry, outlined the prevailing sensors, software, and data analysis techniques, and expounded upon the growing accessibility of diverse data sources.

Wagner et al. [25] conducted research exploring the application of ML for the identification of abnormal behavior through continuous monitoring, utilizing a ruminal bolus to measure pH and predict occurrences of Subacute Ruminal Acidosis (SARA) in cows. In their study, they employed ML techniques, including Decision Tree for Regression (DTR), Multilayer Perceptron (MLP), K-Nearest Neighbors for Regression (KNNR), and Long Short-Term Memory (LSTM). The results indicated that KNNR achieved the highest performance, successfully detecting 83% of SARA incidents.

Unold et al. [26] introduced a Cow Health Monitoring System focused on categorizing the behavior of dairy cows, specifically targeting estrus detection. The study employed a Cow Health Monitoring System composed of hardware components, a cloud system, and an end-user application. Using classification tree analysis, significant features were identified, and the C4.5 decision tree learning algorithm was applied to detect essential behaviors of dairy cows. The developed system demonstrated effectiveness and precision in monitoring dairy cow behavior to identify specific physiological states like estrus or health issues such as mastitis.

Contla Hernández et al. [27] employed mid-infrared (MIR) spectrometry to analyze the content of cow milk. The research explored the feasibility of identifying cow illnesses using Random Forest, Support Vector Machine, Convolutional Neural Network, Neural Network, and ensemble models. The outcomes revealed that employing a neural network yielded a satisfactory level of accuracy in detecting health concerns.

Suseendran and Balaganesh [28] developed a Smart Cattle Health Monitoring System utilizing IoT sensors to effectively anticipate cattle diseases ahead of time. In order to ensure dependable data transmission between the gateway and cow collar, the study introduced a health monitoring module based on fuzzy logic. The system they proposed demonstrated superior performance compared to existing systems, exhibiting a 4% improvement in packet delivery ratio, a 12% enhancement in residual energy, and a 14% reduction in latency, as indicated by the experimental data.

As the domain of cattle health monitoring systems based on IoT continues to garner attention, a consolidated overview of pertinent research papers becomes crucial. In an effort to distill the advancements, methodologies, and findings in this field, a tabular summary of key research works has been curated in Table 1.1 [29]. This tabular exposition not only

TABLE 1.1

Summary of Research Papers on the Application of IoT for Livestock Health Monitoring

References	Objective of the Study	Parameters Observed	Hardware Used	Observations
Swain et al. [15]	To present a device to monitor the health parameters of the cattle to spot any deterioration in the cattle's health	Heart rate, temperature, rumination, and body humidity	Arduino NANO, Xbee module, and a combination of sensors	Accuracy (72%–75%) is not as high because of the simplicity of the developed system.
Benjamin and Yik [21]	To introduce ML algorithms and literature on relevant sensors drawing from industry pig welfare audit criteria and explain how these applications can be used to improve swine welfare	Sound, heat, and motion	Remote sensors such as cameras, microphones, thermometers, and accelerometers	PLF can provide producers with information about the welfare of the whole herd as well as individual animals by collecting and analyzing vast quantities of data.
Eckelkamp [5]	To present the role of wearable precision dairy technologies (WPDT) in the detection of lameness, mastitis, metabolic disorders, and metritis	Feed bunk, rumination time, eating time, lying time, standing time, walking time, activity, lying-to-standing transitions, temperature, and rumen pH	Wearable precision technologies such as neck collars, torso bands, and leg tags	A number of wearable technologies are available to detect diseases in animals, but stronger sensitivity and specificity are needed.
Pratama et al. [20]	To produce normal, less normal, and abnormal health classification outputs based on data collected using sensors and classified using ML	Body temperature, heart rate, and movement	Collar device to read body temperature, heart rate sensor, and motion sensor, Raspberry Pi, a base station for local server management and web application to visualize data and analytic health conditions	The prediction system that is made can make the health of the heart rate and body temperature to be normal, which is less normal and abnormal, with an accuracy of 84% Support Vector Machine (SVM) and using Decision Tree (DT) and 91% accuracy.

Reference	Objective	Parameters	Hardware	Findings
Priya and Jayaram [3]	To describe unique sensors that are associated with wireless sensor networks to monitor the health status of cattle	Temperature, motion and heart rate	Sensor modules, RL78, GSM, and Android application	A framework to monitor an animal's health has been proposed.
Vyas et al. [22]	To reflect on the detection of foot and mouth disease and mastitis disease in cows using of things (IoT)	Temperature, motion, sound, etc.	Rumination sensor, motion sensor, temperature sensor, and Raspberry Pi3 controller	Use of the IoT in the detection of mastitis and FMD reduces the lousy quality of milk and infertility among cows.
Chaudhry et al. [2]	To review the existing technology-based solutions and related equipment and provide a comparison of the features offered by these systems and their limitations	Skin temperature, rumination time, and heart rate its variability	IoT-based sensor unit to measure various health-related parameters	ML can be applied to detect lameness in cattle, Linear Regression Model to predict milk yield, and computer vision and deep learning can be used to analyze health factors.
Unold et al. [26]	To present an IoT-based livestock monitoring system to automate the measurement of dairy cow health state	Feeding, ruminating, resting, and movement	Cow device consisting of sensors, the hub, Wi-Fi access points, and routers, database, and application server	Accurately monitors the behavior of the dairy cows and detects some health problems (e.g., mastitis, estrus).
Neethirajan [4]	To explore the role of sensors, big data, artificial intelligence (AI) and ML to lower production costs, increase efficiencies, enhance animal welfare	Voice signals of animals, visual data of various animal movements, and other such animal behavior data	Hardware sensors such as camera or vision sensors, infrared thermal imaging sensors, temperature sensors, RFID tags, accelerometers, motion sensors, pedometers, facial recognition machine vision sensors, and microphones	Advanced AI and ML algorithms can be applied for the detection of diseases and the monitoring of animals.
Suseendran and Balaganesh [28]	To accurately predict the illness of cattle using a fuzzy-based Smart Cattle Health Monitoring System employing IoT sensors	Heartbeat, body temperature, pulse rate, rumination, and respiration information	Sensors, Wi-Fi module, Arduino	The proposed system achieves a 4% higher packet delivery ratio, 14% lesser delay, and 12% higher residual energy as compared to the existing systems.

underscores the spectrum of perspectives but also elucidates the evolving landscape of IoT-enabled cattle health monitoring.

In summation, the tabular summary presented in Table 1.1 encapsulates a diverse array of research papers integral to the realm of cattle health monitoring systems based on IoT. This compilation not only accentuates the breadth of research methodologies employed but also underscores the pivotal role of IoT technology in modern livestock farming. The amalgamation of these insights not only enriches our comprehension of IoT applications in cattle health but also lays a foundation for more nuanced inquiries into this burgeoning field.

1.5 RESEARCH GAPS

The literature review has uncovered the following research gap within the current systems:

i. *Farmers' Lack of Awareness about Diseased Cattle*: Due to the absence of timely notifications regarding unusual changes in cattle health, farmers often remain unaware of any potential afflictions. This lack of awareness hinders their ability to administer timely treatment, potentially exacerbating the severity of diseases.

ii. *Need for Comprehensive Datasets*: Another notable research gap pertains to the demand for substantial datasets. The utilization of ML techniques for data prediction across various aspects of dairy farming has been observed in existing systems. However, the accuracy of the generated algorithms is contingent upon the availability of extensive, integrated datasets. By analyzing sizable, comprehensive datasets, it is conceivable to enhance decision support systems for farmers, thereby enabling them to advance the welfare and productivity of their livestock.

iii. *Enhancement of ML Algorithm Classification Performance*: A significant research gap pertains to the need for enhancing the classification performance of ML algorithms. Commercially available sensors capable of tracking the health and physiology of dairy cows are in existence. However, there exists ambiguity regarding the effectiveness of the decision support derived from these sensors due to the limited availability of research focusing on the efficacy of the underlying algorithms. This underscores the challenge of improving and predicting the categorization performance of algorithms within the realm of PLF. While the potential of ML has been widely acknowledged by researchers, it remains crucial to collaboratively employ these potent tools alongside data scientists and the dairy industry to comprehensively grasp their true significance.

1.6 COMPARATIVE ANALYSIS

Within the domain of IoT-based cattle health monitoring systems, a diverse array of research studies has surfaced, each contributing its own unique insights and advancements. To distill this wealth of knowledge and provide a structured assessment of these studies, we present a comprehensive comparative analysis. The purpose of this analysis is to offer readers a systematic overview of key research papers, shedding light on their individual strengths and limitations in the quest to enhance cattle health through IoT technology. Our aim is to synthesize these findings, helping readers gain a deeper understanding of the evolving landscape of IoT-enabled cattle health monitoring and identifying promising avenues for future research and development. In Table 1.2, we present a tabular summary of these research works for the convenience of reference and evaluation.

1.7 FUTURE RESEARCH DIRECTIONS

Following the aforementioned comprehensive literature review, it is evident that the landscape of IoT-based cattle health monitoring systems offers immense potential while also revealing areas that require further exploration and innovation. This section delineates the future research directions within the domain of IoT-based cattle health monitoring systems.

 i. *Integration of Cutting-Edge Sensors*: One promising avenue for future research entails the incorporation of advanced sensors and technologies to elevate the precision and breadth of data collection related to cattle health. Delving into the integration of wearable devices, smart collars, and ingestible sensors holds the promise of a more holistic understanding of individual animal health.

TABLE 1.2

Comparative Analysis of Research Papers on IoT-Based Livestock Health
Monitoring Systems

Research Paper	Strengths	Weaknesses
Swain et al. [15]	• Comprehensive sensor usage (humidity, temperature, pulse, rumination) Achieved 72%–75% accuracy in detecting cow health	• Use of older software (Arduino 1.0.6) • Limited exploration of ML techniques
Borchers et al. [16]	• Automated activity, lying, rumination monitors • Explored ML (linear discriminant analysis, Random Forest, neural networks)	• Primarily focused on calving prediction • Limited discussion on data accessibility and management
Liakos et al. [17]	• Holistic approach with various health indicators • Promising results with Random Forest	• Predominantly focused on lameness prediction • Limited discussion on real-time data processing and data accessibility
Cowton et al. [18]	• Utilized deep learning for respiratory illness detection • Rapid response, detecting illness within 1–7 days	• Focused only on pigs • Lack of emphasis on data accessibility and management
Kumari and Yadav [19]	• Cloud-based database for remote data access • Investigated key cattle health indicators	• Limited discussion on ML for data analysis • Scalability impacted by Raspberry Pi 3 as the central controller
Pratama et al. [20]	• Collar devices and ML for health classification • Web application for data display and analysis	• Lack of details on ML techniques and accuracy rates • Limited exploration of sensor types
Benjamin and Yik [21]	• Comprehensive literature review on pig health with IoT • Discussed potential applications for enhancing animal health	• No original research or experiments conducted • No focus on specific sensor technologies or ML techniques
Vyas et al. [22]	• Real-world applications with IoT for disease detection in cows • Use of ML for disorder detection	• Limited discussion on data management and accessibility • Focus primarily on specific diseases, potentially missing broader health monitoring applications

ii. *Predictive Analytics and ML*: The development and application of ML algorithms for predictive analytics in cattle health monitoring represent a pivotal direction. Leveraging historical data, these algorithms can forecast health issues, enabling proactive intervention. However, refining these models to improve accuracy and adaptability across diverse cattle breeds and environmental conditions remains a significant challenge demanding attention.

iii. *Scalability and Cost-Effectiveness*: Scalability remains a paramount concern, especially for small-scale farmers. The future trajectory of research should focus on exploring cost-effective solutions that can be seamlessly adopted across diverse farming operations. This may encompass the investigation of low-cost sensor alternatives, efficient communication protocols, or cloud-based solutions accommodating varying resource constraints.

iv. *User-Friendly Interfaces and Mobile Applications*: User adoption and engagement are pivotal for the success of IoT-based cattle health monitoring systems. The research landscape should prioritize enhancements in user interfaces and the development of mobile applications that facilitate remote monitoring and data management. These efforts aim to render the system accessible and user-friendly for farmers and veterinarians.

In conclusion, the future directions signify opportunities to advance the domain of IoT-based cattle health monitoring systems. Exploring these research domains will not only contribute to livestock welfare and the sustainability of cattle farming practices but also foster the integration of state-of-the-art technology into agriculture, rendering the industry more efficient and productive.

1.8 DISCUSSIONS

Through a comprehensive review of prior research endeavors in this domain, it has become evident that there is a pressing need for an automated system capable of real-time monitoring of animal health. To address this need, we propose an IoT-based solution model aimed at detecting diseases in cows within the context of PLF. This proposed system is designed to offer dairy farmers timely alerts or notifications concerning cow diseases, facilitated by the collection of behavioral data through real-time sensor monitoring. This proactive approach not only aids in averting disease outbreaks but also minimizes potential losses incurred as a result.

To effectively employ mechanistic models in dairy farming, the acquisition of extensive and diverse datasets becomes essential. These datasets can be efficiently gathered through the utilization of various sensors. Predicting diseases in dairy cows can be achieved by implementing an IoT-based hardware device and leveraging ML algorithms to discern the health status of the cows. The ensuing steps required for this process are succinctly outlined as follows:

   i. *Data Collection Using Sensors*: Real-time data from animals, such as cows, is collected using an array of hardware sensors. These sensors include accelerometers, motion sensors, pulse rate sensors, temperature sensors, and microphones. This diverse set of sensors helps gather comprehensive information about the animals' behavior, movement, health indicators, and environmental conditions.
  ii. *Data Transmission to Raspberry Pi*: The data collected by the sensors is then transmitted to a Raspberry Pi microcontroller. This microcontroller is equipped with a built-in Wi-Fi module that enables the transfer of the sensed data to a cloud-based platform or a designated server for further analysis.
 iii. *Advanced ML Analysis*: The broad dataset obtained from the sensors is then subjected to analysis using advanced ML algorithms. These algorithms leverage the collected information to analyze, predict, and provide alerts to farmers in case any abnormalities or unusual patterns are detected. The algorithms can perform tasks like clustering, classification, and pattern recognition, which are crucial for monitoring and identifying potential diseases in animals through sophisticated data analysis techniques.
  iv. *Alerts via Mobile App*: The determined health status of the cows is communicated to the farmer through alerts sent via a dedicated mobile application. These alerts inform the farmer about the detected conditions of the cows, allowing for timely intervention and necessary actions to ensure animal welfare.

In essence, this process involves the integration of sensors, data transmission, advanced ML, and real-time alerts to empower farmers with the means to proactively manage their livestock's health and well-being.

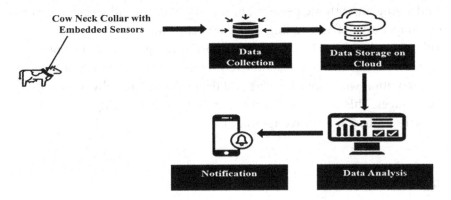

FIGURE 1.1

Proposed system architecture to predict diseases in dairy cows.

As a result, when it comes to cattle farming, sensors, big data, advanced AI, and ML are all intertwined to provide a comprehensive solution.

Figure 1.1 diagrammatically represents the proposed system architecture.

1.9 CONCLUSION

Based on an analysis of existing research on disease detection in livestock, several key conclusions can be drawn. First, cattle are susceptible to a variety of diseases. Unfortunately, the absence of notifications when abnormal health variations occur means that farmers remain unaware of potential illnesses affecting their cattle. Consequently, farmers may not provide timely treatment, leading to the potential exacerbation of disease severity. Examining the landscape of current systems reveals that ML methodologies exhibit predictive capabilities across various domains within dairy farming. However, the reliability of these algorithms hinges on the availability of extensive integrated datasets. The potential for farmers to enhance their decision support systems through the analysis of these comprehensive datasets could greatly improve the well-being and efficiency of their livestock. The availability of commercial sensors designed to monitor dairy cow physiology and health indicators is noteworthy. Nonetheless, the efficacy of the decision assistance provided by these sensors remains unclear due to the scarcity of published data concerning the underlying algorithmic performance. A significant challenge pertains to the imperative of forecasting and enhancing the performance of algorithms used within PLF. Despite widespread recognition of the potential of ML, it is essential to collaboratively implement these powerful tools in collaboration with dairy farmers and data scientists to fully unlock their advantages. This collective effort is crucial to fully realizing the potential benefits of these innovative approaches.

REFERENCES

1 Banhazi, T. M., Lehr, H., Black, J. L., Crabtree, H., Schofield, P., Tscharke, M., & Berckmans, D. (2012). Precision livestock farming: an international review of scientific and commercial aspects. *International Journal of Agricultural and Biological Engineering*, 5(3), 1–9.

2 Chaudhry, A. A., Mumtaz, R., Zaidi, S. M. H., Tahir, M. A., & School, S. H. M. (2020, December). Internet of Things (IoT) and machine learning (ML) enabled livestock monitoring. In *2020 IEEE 17th International Conference on Smart Communities: Improving Quality of Life Using ICT, IoT and AI (HONET)* (pp. 151–155). IEEE.

3 Priya, M. K., & Jayaram, B. G. (2019). WSN-based electronic livestock of dairy cattle and physical parameters monitoring. In *Emerging Research in Electronics, Computer Science and Technology* (pp. 37–45). Springer, Singapore.

4 Neethirajan, S. (2020). The role of sensors, big data and machine learning in modern animal farming. *Sensing and Bio-Sensing Research*, 37, 100367.

5 Eckelkamp, E. A. (2019). Invited review: current state of wearable precision dairy technologies in disease detection. *Applied Animal Science*, 35(2), 209–220.

6 Bewley, J. M., Borchers, M. R., Dolecheck, K. A., Lee, A. R., Stone, A. E., & Truman, C. M. (2017). Precision dairy monitoring technology implementation opportunities and challenges. *Large Dairy Herd Management*, 3, 1251–1264.

7 Halachmi, I., Guarino, M., Bewley, J., & Pastell, M. (2019). Smart animal agriculture: application of real-time sensors to improve animal well-being and production. *Annual Review of Animal Biosciences*, 7, 403–425.

8 Vranken, E., & Berckmans, D. (2017). Precision livestock farming for pigs. *Animal Frontiers*, 7(1), 32–37.

9 Berckmans, D. (2017). General introduction to precision livestock farming. *Animal Frontiers*, 7(1), 6–11.

10 Schillings, J., Bennett, R., & Rose, D. C. (2021). Exploring the potential of precision livestock farming technologies to help address farm animal welfare. *Frontiers in Animal Science*, 2.

11 Berckmans, D. (2014). Precision livestock farming technologies for welfare management in intensive livestock systems. *Revue scientifique et technique*, 33(1), 189–196.

12 Pomar, J., López, V., & Pomar, C. (2011). Agent-based simulation framework for virtual prototyping of advanced livestock precision feeding systems. *Computers and Electronics in Agriculture*, 78(1), 88–97.

13 Cockburn, M. (2020). Application and prospective discussion of machine learning for the management of dairy farms. *Animals*, 10(9), 1690.

14 Hyde, R. M., Down, P. M., Bradley, A. J., Breen, J. E., Hudson, C., Leach, K. A., & Green, M. J. (2020). Automated prediction of mastitis infection patterns in dairy herds using machine learning. *Scientific Reports*, 10(1), 1–8.

15 Swain, K. B., Mahato, S., Patro, M., & Pattnayak, S. K. (2017, May). Cattle health monitoring system using Arduino and LabVIEW for early detection of diseases. In *2017 Third International Conference on Sensing, Signal Processing and Security (ICSSS)* (pp. 79–82). IEEE.

16 Borchers, M. R., Chang, Y. M., Proudfoot, K. L., Wadsworth, B. A., Stone, A. E., & Bewley, J. M. (2017). Machine-learning-based calving prediction from activity, lying, and ruminating behaviors in dairy cattle. *Journal of Dairy Science*, 100(7), 5664–5674.

17 Liakos, K., Moustakidis, S. P., Tsiotra, G., Bartzanas, T., Bochtis, D., & Parisses, C. (2017). Machine Learning based computational analysis method for cattle lameness prediction. In *HAICTA* (pp. 128–139).

18 Cowton, J., Kyriazakis, I., Plötz, T., & Bacardit, J. (2018). A combined deep learning gru-autoencoder for the early detection of respiratory disease in pigs using multiple environmental sensors. *Sensors*, 18(8), 2521.

19 Kumari, S., & Yadav, S. K. (2018). Development of IoT based smart animal health monitoring system using Raspberry Pi. *International Journal of Advanced Studies of Scientific Research*, 3(8), 24–31.

20 Pratama, Y. P., Basuki, D. K., Sukaridhoto, S., Yusuf, A. A., Yulianus, H., Faruq, F., & Putra, F. B. (2019, September). Designing of a smart collar for dairy cow behavior monitoring with application monitoring in microservices and internet of things-based systems. In *2019 International Electronics Symposium (IES)* (pp. 527–533). IEEE.

21 Benjamin, M., & Yik, S. (2019). Precision livestock farming in swine welfare: a review for swine practitioners. *Animals*, 9(4), 133.

22 Vyas, S., Shukla, V., & Doshi, N. (2019). FMD and mastitis disease detection in cows using Internet of Things (IOT). *Procedia Computer Science*, 160, 728–733.

23 Xu, W., van Knegsel, A. T., Vervoort, J. J., Bruckmaier, R. M., van Hoeij, R. J., Kemp, B., & Saccenti, E. (2019). Prediction of metabolic status of dairy cows in early lactation with on-farm cow data and machine learning algorithms. *Journal of Dairy Science*, 102(11), 10186–10201.

24 Garcia, R., Aguilar, J., Toro, M., Pinto, A., & Rodriguez, P. (2020). A systematic literature review on the use of machine learning in precision livestock farming. *Computers and Electronics in Agriculture*, 179, 105826.

25 Wagner, N., Antoine, V., Mialon, M. M., Lardy, R., Silberberg, M., Koko, J., & Veissier, I. (2020). Machine learning to detect behavioural anomalies in dairy cows under subacute ruminal acidosis. *Computers and Electronics in Agriculture*, 170, 105233.

26 Unold, O., Nikodem, M., Piasecki, M., Szyc, K., Maciejewski, H., Bawiec, M., ... & Zdunek, M. (2020, June). IoT-based cow health monitoring system. In *International Conference on Computational Science* (pp. 344–356). Springer, Cham.

27 Contla Hernández, B., Lopez-Villalobos, N., & Vignes, M. (2021). Identifying health status in grazing dairy cows from milk mid-infrared spectroscopy by using machine learning methods. *Animals*, 11(8), 2154.

28 Suseendran, G., & Balaganesh, D. (2021). Smart cattle health monitoring system using IoT sensors. *Materials Today: Proceedings.*

29 Kaur, D., & Kaur, A. (2022). IoT and machine learning-based systems for predicting cattle health status for precision livestock farming. In *2022 International Conference on Smart Generation Computing, Communication and Networking (SMART GENCON)* (pp. 1–5). IEEE.

2

Significance of Machine Learning in Apple Disease Detection and Implications

Saimul Bashir
Chandigarh University, Mohali, India

Syed Nisar Hussain Bukhari
National Institute of Electronics and Information Technology (NIELIT),
Srinagar, India

Faisal Firdous
Jaypee University of Information and Technology, Solan, India

Gursimran Jeet Kour
Chandigarh University, Mohali, India

2.1 INTRODUCTION

Apples are one of the most popular fruits worldwide, with a global production of over 80 million tons in 2020 [1]. However, apple crops are vulnerable to a wide range of diseases, which can lead to significant economic losses for growers. Common apple diseases include apple scab, powdery mildew, fire blight, and cedar apple rust, which are caused by fungi, bacteria, and viruses. These diseases can affect the quality and quantity of fruit produced, leading to lower yields, increased costs for growers, and reduced consumer demand [2].

DOI: 10.1201/9781003485179-2

2.1.1 Brief Description of Traditional Disease Detection Methods

Traditionally, apple disease detection has relied on visual inspection of crops by experts, which can be time-consuming and expensive [3]. This approach is also subjective and can lead to inconsistent results. Other methods, such as laboratory tests and chemical sprays, are also used but have their own limitations [4]. For example, laboratory tests can be expensive and time-consuming, while chemical sprays can be harmful to the environment and human health. The six disease types are given in Figure 2.1.

2.1.2 Importance of Machine Learning in Disease Detection

In several industries, including healthcare, finance, and resource management, machine learning (ML) has become a potent tool for illness identification and management. This technology uses advanced algorithms and statistical models to analyze large datasets and identify patterns and correlations that would be difficult or impossible to detect through manual analysis [5]. By doing so, ML can help to improve the accuracy and efficiency of disease detection, reduce the risk of misdiagnosis and unnecessary treatments, and ultimately improve patient outcomes.

ML may be applied in agriculture to enhance disease detection and management effectiveness and efficiency while lowering the environmental

FIGURE 2.1
Apple leaf disease types.

impact of agricultural practices and enhancing the sustainability and profitability of farming operations [6]. In particular, ML models can be trained to analyze large datasets of images, sensor readings, and other types of data to identify the presence and severity of diseases in crops like apples [2].

The capability of ML to learn and adapt over time makes it one of the most useful features for illness diagnosis. As the algorithm is exposed to more data, it can improve its accuracy and become more effective at identifying subtle patterns and correlations [7]. This can be particularly useful in the context of apple disease detection, where new and emerging diseases may be difficult to identify using traditional methods. We can enhance our capacity to identify and respond to these new dangers by utilizing the power of ML, therefore defending the well-being and production of apple harvests [3]. The capacity of ML to accurately and quickly analyze massive volumes of data is another crucial advantage [8]. In the context of agriculture, this can be particularly important, as the timely detection and management of diseases can have a significant impact on crop yields and profitability. By using ML models to analyze large datasets of images, sensor readings, and other types of data, we can quickly and accurately identify the presence and severity of diseases in crops like apples, allowing for more timely and effective management strategies [9].

Moreover, ML can also help to reduce the need for human labor in disease detection (Figure 2.2) and management, reducing costs and increasing efficiency [7]. By automating the process of disease detection, we can reduce the need for manual labor and free up resources for other tasks.

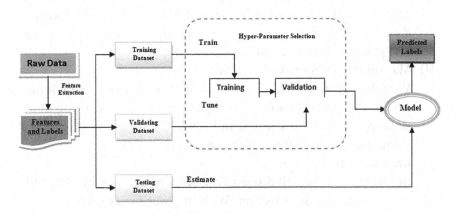

FIGURE 2.2

Generalized diagram of ML for disease detection.

This can be particularly beneficial in the context of rural and remote farming communities, where labor is often scarce and costly [10].

2.1.3 Objectives of the Chapter

This chapter's primary goals are to present an overview of the application of ML for the diagnosis of apple disease, to review the literature on this topic, and to discuss the strengths and limitations of ML in this context. The chapter will also cover the dataset preparation, ML model selection, and evaluation techniques used in apple disease detection, as well as future directions for research in this area. In recent years, there has been a growing interest in the use of ML for disease detection in agriculture. ML algorithms have shown promise in detecting diseases in crops such as potatoes, wheat, and grapes. However, the use of ML for apple disease detection is still in its early stages. This is partly due to the complex nature of apple diseases, which can be caused by multiple factors, including environmental conditions, genetic factors, and pathogens.

The chapter is organized as follows:

I. Related Work

In this section, we will provide a brief review of the existing state-of-art on apple disease detection using several ML techniques. We will try to highlight the performance and limitations of existing techniques in this section.

II. Dataset Preparation

The significance of dataset preparation in ML for apple disease detection will be covered in this section. We will cover topics such as data collection, data cleaning, and data augmentation, and discuss best practices for preparing datasets for ML.

III. Machine Learning Model Selection

We will examine the research on ML models for detecting apple disease in this part. We will cover topics such as deep-learning models, support vector machines, and random forests, and discuss the strengths and weaknesses of each approach.

IV. Evaluation Techniques

In this section, we will discuss evaluation techniques for ML models in apple disease detection. We will cover topics such as cross-validation, precision-recall curves, and confusion matrices, and discuss best practices for evaluating ML models in this context.

V. Ethical, Practical, and Technical Implications/Considerations

In this section, we will discuss some of the important ethical, practical, and technical implications of using ML models for apple disease detection. As with any technology, there is a risk of unintended consequences in developing these models.

VI. Strengths and Limitations of Machine Learning

This section will discuss the advantages and disadvantages of ML for identifying apple illnesses. We'll talk about things like the risk of overfitting, the value of high-quality datasets, and the possibility of data bias.

VII. Future Directions for Research

In this section, we will identify future directions for research in ML for apple disease detection. We will discuss topics such as the use of multimodal data, the development of transfer learning models, and the use of unsupervised learning techniques.

VIII. Conclusion

We will highlight the chapter's key ideas and talk about the implications of ML for identifying apple diseases in this part. We'll examine the possible effects of ML on the apple business as well as areas that need more study.

2.2 RELATED WORK

A multiclass support vector machine (SVM) technique was put up by Soarov Chakraborty et al. [11] to categorize apple leaves into three groups: healthy, scabbed, and powdery mildew. The dataset used to train and test the system is thoroughly described in the publication, along with the methods used to acquire the images, preprocess them, and extract their features. The implementation of the SVM algorithm and the performance measures employed to assess the method are then described by the authors. The outcomes demonstrate that the suggested multiclass SVM algorithm classified apple leaves into healthy, scab, and powdery mildew classes with good accuracy rates. The limits of the suggested technique are discussed in the paper's conclusion, along with ideas for more research.

A convolutional neural network (CNN) model by Swati et al. [12] was reported that can divide apple leaf pictures into four categories: healthy, scab, rust, and powdery mildew. The CNN model's implementation and the performance measures employed to assess the model's correctness are both described by the author. The outcomes demonstrate that the proposed CNN model classified apple leaf pictures into several disease groups with good accuracy rates. The limits of the suggested technique are discussed in the paper's conclusion, along with ideas for more research. Overall, the research proposes a well-designed method for employing deep learning to automatically detect apple leaf illnesses. An interesting method that has the potential to be extended to other plant species and might have practical ramifications for agriculture is the use of a CNN model to categorize the photos into several disease groups. Xin li et al. [13] propose a deep-learning approach to detect and classify apple leaf diseases. The authors use residual neural network (ResNet) models to classify apple leaf images into one of three classes: healthy, scab, and powdery mildew. A thorough explanation of the dataset, which consists of photos gathered from various sources and was used to train and test the ResNet models, is also provided. The authors also go into how ResNet models are put into practice and what performance indicators are used to gauge how accurate the model is. The outcomes demonstrate that the suggested ResNet models successfully classified apple leaf pictures into various disease classes with high accuracy rates. The limits of the suggested technique are discussed in the paper's conclusion, along with ideas for more research. An intriguing method that has the potential to be extended to other plant species and might have practical ramifications for agriculture is the use of ResNet models to categorize the photos into several disease groups. Saraansh Baranwal et al. [14] propose a deep-learning approach for detecting apple leaf diseases using CNNs. The author presents a model that can classify apple leaf images into one of four classes: healthy, scab, rust, and powdery mildew. The CNN model's implementation and the performance measures employed to assess the model's correctness are both described by the author. The outcomes demonstrate that the proposed CNN model classified apple leaf pictures into several disease groups with good accuracy rates. A hybrid strategy is suggested by Mohammed Khalid Kaleem et al. [15] for the automated identification of leaf diseases. The author employs SVM classifiers to categorize the pictures as healthy or ill after using image processing techniques to preprocess the images and extract the features. On photos of numerous plant species obtained from diverse sources, the SVM classifier was trained and tested. The author also discusses the SVM classifier's implementation and the performance measures employed to gauge the model's precision. The outcomes demonstrate that the suggested method successfully classified leaf pictures into healthy and sick at high rates of accuracy.

With the installation of the CNN model and the performance measures used to assess the model's correctness, Srinidhi et al. [16] classified apple leaf pictures into one of four classes: healthy, apple scab, cedar apple rust, and quince rust. The outcomes demonstrate that the proposed CNN model successfully categorizes apple leaf pictures into several disease classes with high accuracy rates. The limits of the suggested technique are discussed in the paper's conclusion, along with ideas for more research. Overall, the research proposes a well-designed method for employing deep learning to automatically detect apple leaf illnesses. The use of a CNN model to classify the images into different disease classes is a promising approach that has the potential to be applied to other plant species and could have practical implications in the field of agriculture.

Hyperspectral imaging is suggested by Mubarakat Shuaibu et al. [17] as a method for the automated identification of Marssonina blotch disease in apple plants. The most pertinent spectral bands for diagnosing the condition are determined by the author using a feature selection approach based on mutual information. The hyperspectral dataset used for feature selection and the execution of the unsupervised strategy is thoroughly described in the publication. The findings demonstrate that the suggested method successfully detected Marssonina blotch disease in apple trees with high accuracy rates. Overall, the paper presents a novel approach to the automated detection of Marssonina blotch disease in apple trees using hyperspectral imaging and an unsupervised feature selection technique. The use of hyperspectral imaging could provide a noninvasive and efficient way to detect plant diseases, and the proposed approach has the potential to be applied to other plant species and could have practical implications in the field of agriculture.

CNNs are proposed by Prakhar Bansal et al. [18] for the automation of the identification of illnesses in apple leaves. The author offers a methodology that can categorize pictures of apple leaves into three groups: healthy, scabby, and rusted. The study offers a viable method for applying deep learning to automatically detect illnesses in apple leaves. A strong method that has the potential to be extended to other plant species and might have practical ramifications for agriculture is the use of a CNN model to categorize the photos into several disease groups.

SVMs are an ML technique proposed by S. Manimegalai et al. [19] for apple leaf disease detection. The author also discusses the SVM model's implementation and the performance measures employed to gauge the model's precision. The outcomes demonstrate that the suggested SVM model classified apple leaf pictures into several illness classes with good accuracy rates. An effective method that has the potential to be extended to other plant species and might have practical ramifications for agriculture is the use of an SVM model to categorize the photos into several disease classifications.

A method for image processing is suggested by Sara Alqethami et al. [20] for the automated identification of illnesses in apple leaves. The author describes an approach that starts with picture preprocessing, segmentation, and feature extraction, then uses ML algorithms to classify the images. The dataset used to train and evaluate the classification model, which contains photos obtained from several Saudi Arabian orchards, is described in depth in the publication. The author also discusses the application of the classification model utilizing the decision tree, SVM, and k-nearest neighbors ML techniques. The outcomes demonstrate that the suggested method classified apple leaf photos into several disease groups with good accuracy rates.

2.3 DATASET PREPARATION

In ML, the quality of the dataset is crucial for the performance of the model. Therefore, data preparation is an important step in the ML pipeline. In the context of apple disease detection, data preparation involves collecting data, cleaning it, and augmenting it to increase the diversity of the dataset.

2.3.1 Data Collection

Collecting high-quality data is the first step in preparing a dataset for ML. In the context of apple disease detection, data can be collected using different techniques such as manual sampling, remote sensing, and drones. Manual sampling involves physically examining the apple trees and recording the symptoms of any diseases observed. Remote sensing involves using aerial imagery or satellite imagery to detect changes in the vegetation of the orchard, which can be indicative of disease. Drones can also be used to collect high-resolution images of the orchard and detect changes in the vegetation.

2.3.2 Data Cleaning

After collecting the data, the next step is to clean it. Data cleaning involves removing outliers, correcting errors, and handling missing values. In the context of apple disease detection, this can involve removing images that are of poor quality or contain objects that are not relevant to the task, such as images of birds or clouds. It can also involve correcting labeling errors, such as mislabeling a healthy tree as diseased, and handling missing values, such as missing annotations for certain images.

2.3.3 Data Augmentation

Data augmentation is the process of artificially increasing the diversity of the dataset by applying transformations such as rotation, flipping, and cropping to the original images. Data augmentation can help to improve the robustness of the model by exposing it to variations in the input data. In the context of apple disease detection, data augmentation can involve applying transformations to the original images, such as rotating them or flipping them, and adding noise or blurring effects.

2.4 MACHINE LEARNING MODEL SELECTION

We will examine the research on ML models for detecting apple disease in this part. Deep-learning models, SVMs, and random forests are just a few of the subjects we'll explore, along with each method's advantages and disadvantages.

2.4.1 Deep-Learning Models

Modern performance has been attained by deep-learning models in a variety of computer vision applications, including segmentation, object identification, and picture classification [21]. Deep-learning models have demonstrated excellent results in the area of apple disease detection, reaching high accuracy in identifying various illnesses. One of the most commonly used deep-learning models for apple disease detection is CNN (Figure 2.3) [22]. CNNs are a type of neural network designed to process images, and they have achieved state-of-the-art performance in many

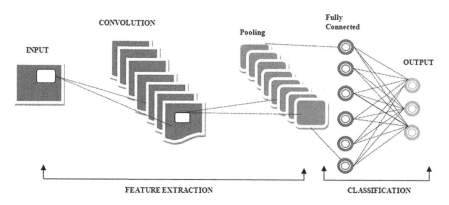

FIGURE 2.3

CNN working [9].

computer vision tasks. In the context of apple disease detection, CNNs have been used to detect diseases such as apple scab, powdery mildew, and fire blight [23].

Another type of deep-learning model that has been used for apple disease detection is the recurrent neural network (RNN). RNNs are a type of neural network designed to process sequential data, such as time series or text. In the context of apple disease detection, RNNs have been used to detect diseases based on the temporal patterns of symptoms, such as changes in color or texture [24].

2.4.2 Support Vector Machines

SVMs (Figure 2.4) are a type of supervised learning algorithm that can be used for both classification and regression tasks [15]. SVMs are particularly well-suited to classification tasks in which the data is not linearly separable, as they can map the data to a higher-dimensional space where it is linearly separable [25]. SVMs have been used to categorize photos of healthy and sick apple leaves in the context of detecting apple diseases. Hyperspectral SVMs have also been used to find apple scab and powdery mildew [26].

2.4.3 Random Forests

Multiple decision trees are combined in random forests, an ensemble learning technique, to increase the model's robustness and accuracy [27]. When making a forecast, random forests (Figure 2.5) combine the results of all the decision trees they created using different random subsets of

Training (offline)

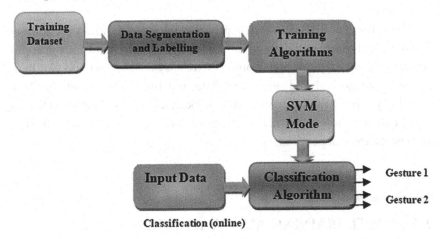

FIGURE 2.4

SVM working.

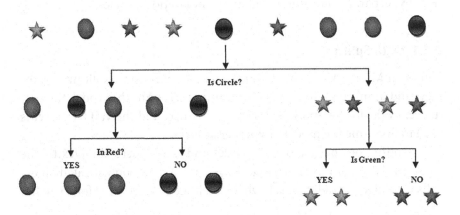

FIGURE 2.5

Random forest diagram.

the characteristics and data [8]. In the context of apple disease detection, random forests have been used to classify images of healthy and diseased apple leaves. Random forests have also been used to detect diseases such as apple scab and powdery mildew using spectral data [28].

2.4.4 Other Machine Learning Models

In addition to deep-learning models, SVMs, and random forests, there are many other ML models that can be used for apple disease detection [29].

These include decision trees, Naive Bayes, k-nearest neighbors, and logistic regression.

Each of these models has its strengths and weaknesses, and the choice of which model to use will depend on the specific requirements of the application [30]. For example, decision trees are easy to interpret and can handle both categorical and numerical data, but they may overfit the data. Naive Bayes is a straightforward and quick model that performs well with small datasets; however, it makes the unrealistic assumption that all characteristics are independent.

2.5 MODEL TRAINING AND EVALUATION

In this section, we will discuss the process of training and evaluating ML models for apple disease detection. We will cover topics such as data splitting, hyperparameter tuning, and model evaluation metrics.

2.5.1 Data Splitting

Before an ML model is trained on the dataset, it must be split into training, validation, and testing sets. The training set trains the model, the testing set evaluates the model's final performance, and the validation set is used to fine-tune the model's hyperparameters.

Depending on the size of the dataset and the model's complexity, the dataset's splitting ratio can change. In general, a more sophisticated model calls for a bigger validation set, whereas a larger dataset calls for a smaller one.

2.5.2 Hyperparameter Tuning

ML models have hyperparameters that need to be set before training, such as the learning rate, regularization strength, and number of layers. Hyperparameter tuning involves finding the optimal values for these hyperparameters to improve the performance of the model [38].

There are different methods for hyperparameter tuning, such as grid search, random search, and Bayesian optimization. Grid search involves evaluating the model for all possible combinations of hyperparameters, while random search involves randomly selecting hyperparameters to evaluate. Bayesian optimization uses a probabilistic model to select the next set of hyperparameters to evaluate based on the previous results.

2.5.3 Model Evaluation Metrics

Metrics for measuring the performance of the model are utilized (Table 2.1). Accuracy, precision, recall [30], and F1 score are typical assessment criteria in the context of apple disease detection [15].

The percentage of samples in the testing set that were properly categorized is considered the accuracy. Out of all the samples that were expected to be positive, precision is the proportion of genuine positives. Recall quantifies the proportion of real positive samples among all positive samples. The F1 score, which offers a single number to assess the overall effectiveness of the model, is the harmonic mean of accuracy and recall [30].

2.6 ETHICAL, PRACTICAL, AND TECHNICAL CONSIDERATIONS/IMPLICATIONS

It's crucial to take into account the ethical ramifications of utilizing ML algorithms to identify apple diseases. The data and algorithms used to create these models have the same risks as any other technology in terms of unexpected outcomes or bias.

TABLE 2.1

Different Performance Evaluation Metrics

Model Evaluation Metrics	Formulas
Accuracy	(TP + TN) / (TP + TN + FP + FN) [15]
Precision	TP / (TP + FP) [15]
Recall	TP / (TP + FN) [15]
F1 Score	2 * (Precision * Recall)/(Precision + Recall) [30]
Fβ Score	$(1 + \beta^2)$ * (Precision * Recall)/(β^2 * Precision + Recall) [30]
ROC-AUC Score	area under the ROC curve [16]
Log Loss	$-\frac{1}{n}\Sigma i \Sigma j \, yij \log(pij)$ [16]
Mean Squared Error	$-\frac{1}{n}\Sigma i \left(yi - \hat{yi} \right)^2$ [16]

For example, if ML models are developed based on data collected from a single region or climate, they may not perform well when deployed in different regions or under different environmental conditions. This could lead to ineffective disease management strategies or increased use of pesticides, which can have negative environmental and health impacts [31].

Moreover, there is also a risk of bias in the data used to develop these models. If the data used to train the model is biased toward certain demographics or geographic regions, this can lead to biased predictions and potentially discriminatory outcomes. It is, therefore, important to ensure that the data used to develop these models is diverse and representative of the populations and environments in which they will be deployed.

Another ethical consideration is the potential impact of ML models on the livelihoods of farmers and other agricultural workers. If ML models are used to automate disease detection and management tasks, this could lead to job loss or reduced job security for these workers. It is, therefore, important to consider the potential impact of these technologies on the livelihoods of those working in the agricultural sector.

In conclusion, even though ML techniques show tremendous potential for enhancing the efficacy and efficiency of managing apple disease, it is crucial to think about the ethical implications of their usage. It is crucial to make sure that these models are created and implemented in a fair, open, and equitable manner and do not unintentionally harm the environment, the public's health, or agricultural employees. By overcoming these obstacles, we can fully utilize ML for the diagnosis of apple diseases and contribute to the creation of a more just and sustainable agricultural system.

2.6.1 Practical Consideration

Finally, it is important to consider the practical considerations of implementing ML models for apple disease detection in real-world settings. One key consideration is the need for specialized equipment and expertise for data collection and analysis. This may require additional training or investment in equipment and resources, which can be a barrier for smaller farmers or those with limited resources. Another practical consideration is the need for ongoing maintenance and updates to ML models. As new disease strains emerge or environmental conditions change, it may be necessary to update the models to ensure their accuracy and effectiveness [32]. This requires ongoing investment in research and development, as well as a commitment to ongoing monitoring and evaluation of the models in real-world settings. Moreover, it is also important to consider the potential costs and benefits of implementing ML models for apple disease detection [17]. While these models can improve the efficiency and accuracy of disease detection and management, they may also require significant upfront costs for data collection and analysis, as well as ongoing investment in maintenance and updates. It is, therefore, important to carefully consider the potential benefits and costs of these technologies before implementing them on a large scale [22]. In conclusion, the development and deployment of ML models for apple disease detection represents a significant advancement in the field of plant pathology and agriculture. These models have the potential to improve the efficiency and accuracy of disease detection and management while also providing new insights into the underlying biology of plant diseases [33]. However, there are also significant challenges to the development and deployment of these models, including the need for large and diverse datasets, interpretability in ML models, and ethical considerations related to bias and unintended consequences. With continued research and development, we can expect to see significant advancements in this field and the potential for ML to transform the way we manage plant diseases in the agricultural sector.

2.6.2 Technical Considerations

In addition to the challenges and ethical considerations discussed earlier, there are also several technical considerations that must be taken into account when developing ML models for apple disease detection. One key consideration is the need for high-quality and diverse data to train these models [7]. This requires the collection of large amounts of data from different geographic locations, seasons, and apple varieties, as well as the inclusion of data on various environmental factors, such as temperature, humidity, and rainfall.

Another technical consideration is the selection and optimization of ML algorithms for apple disease detection. There are many different algorithms available for use in ML, each with its own strengths and weaknesses [34]. It is therefore important to carefully evaluate and compare different algorithms to determine which ones are most effective for apple disease detection and to optimize these algorithms for maximum accuracy and efficiency.

When creating ML models for the diagnosis of apple disease, interpretability is another crucial factor to take into account. Despite the great levels of accuracy that these models may reach, it can be challenging to comprehend how they make their predictions, especially when deeper learning techniques are used [35]. This lack of interpretability can make it challenging to identify the underlying causes of disease outbreaks and to develop effective management strategies.

Another important factor is making sure that these models are reliable and resilient. ML models are prone to mistakes, especially if they were trained on biased or unreliable data. Therefore, it's crucial to make sure the data used to train these models is impartial and representative and to utilize tools like cross-validation and error analysis to spot and fix any possible mistakes [17].

Finally, there is also a need to ensure that these models are scalable and adaptable to different agricultural contexts and geographies. This requires the development of models that are transferable across different regions and apple varieties, as well as the incorporation of local knowledge and expertise to ensure that these models are relevant and effective in specific agricultural contexts [18].

One promising avenue for the continued development and deployment of ML models for apple disease detection and management is through the use of collaborative research and development initiatives. Such initiatives bring together researchers, farmers, industry partners, and other stakeholders to share knowledge, resources, and expertise and to collaborate on the development of innovative solutions to the challenges facing the agricultural sector.

One such initiative is the Plant Phenotyping and Imaging Research Centre (P2IRC) at the University of Saskatchewan, which is focused on the development of advanced imaging and ML technologies for crop phenotyping, disease detection, and precision agriculture [36]. Another example is the Initiative for Digital Agriculture and Food (IDAF) at the University of Guelph, which is focused on the development and deployment of digital technologies, including ML, for sustainable agriculture and food systems [37]. In addition to these research initiatives, there are also a growing number of industry partnerships and collaborations focused on the development and deployment of ML models for apple disease detection and management. These partnerships bring together technology companies, farmers, and other stakeholders to develop innovative solutions that can be scaled and deployed across different agricultural contexts. For example, the agtech company CropX has developed a soil moisture monitoring system that uses ML algorithms to predict soil moisture levels and help farmers optimize irrigation and fertilizer use. Similarly, the start-up AgShift has developed an ML platform that can analyze the quality and freshness of fruits and vegetables in real time, helping to reduce food waste and improve supply chain efficiency. Overall, the continued development and deployment of ML models for apple disease detection and management represent a significant opportunity to address the challenges facing the agricultural sector today [38–41]. By leveraging the power of data and advanced algorithms, we can improve the accuracy and efficiency of disease detection and management, reduce the environmental impact of agricultural practices, and ultimately improve the sustainability and profitability of farming operations. However, realizing the full potential of these technologies will require continued investment in research and development, as well as a commitment to addressing the technical, ethical, and practical considerations that arise with their use [42–45]. By doing so, we can build a more sustainable and resilient agricultural system for the future and ensure the availability of healthy and nutritious food for generations to come.

2.7 CHALLENGES AND FUTURE DIRECTIONS

In this section, we will discuss some of the challenges and future directions for apple disease detection using ML techniques.

2.7.1 Limited Dataset Availability

One of the main challenges in apple disease detection using ML is the limited availability of high-quality datasets [46–47]. Collecting and annotating data can be time-consuming and expensive, and there is a need for more publicly available datasets to train and evaluate ML models.

2.7.2 Generalization to New Environments

The capacity of the model to generalize to different contexts is another difficulty in apple disease detection using ML. On fresh datasets gathered from other places or under various growth circumstances, ML models trained on a particular dataset may not function effectively. Models that can adapt to changing situations are therefore required.

2.7.3 Interpretable Models

Interpretable ML models are important in the context of apple disease detection, as they can provide insights into the underlying mechanisms of the disease. However, many deep-learning models are not easily interpretable, and there is a need for models that can provide interpretable results.

2.7.4 Incorporating Domain Knowledge

Incorporating domain knowledge into ML models can improve their accuracy and robustness. For example, models that incorporate knowledge of the biology of the apple plant and the disease symptoms can improve the accuracy of disease detection.

2.7.5 Integration with Precision Agriculture

ML models for apple disease detection can be integrated with precision agriculture techniques to improve the efficiency and effectiveness of disease management. For example, models can be used to identify the optimal time for pesticide application or to monitor disease progression in real-time.

The creation and use of ML models for apple disease diagnosis, however, is not without its difficulties. The absence of large-scale, high-quality datasets is a significant problem. ML models must be trained on big, varied datasets that capture the complete spectrum of illness symptoms and environmental variables if they are to function effectively.

The requirement for interpretability in ML models presents another difficulty. Deep-learning models may be highly accurate, but because of their "black-box" character, it is challenging to comprehend the fundamental principles that underlie their predictions. Interpretable models can offer information on the biology of the plant and the illness, which is crucial in the context of disease detection.

Despite these challenges, the development and deployment of ML models for apple disease detection have already shown promising results. In particular, the use of CNNs and other deep-learning models has shown significant improvements in disease detection accuracy compared to traditional image processing techniques.

2.8 CONCLUSION

In this chapter, we have discussed the use of ML techniques for apple disease detection. We have covered the different types of ML models that have been used for this application, the process of model training and evaluation, and the challenges and future directions for this field. Overall, ML has the potential to transform how we identify and treat apple illnesses, and further study and development are required in this field. In the upcoming years, we may anticipate seeing major increases in the precision and effectiveness of apple disease detection due to the expanding availability of high-quality datasets and the development of increasingly advanced ML models. Furthermore, the integration of ML models with precision agriculture techniques can lead to more targeted and efficient disease management strategies, reducing the use of pesticides and improving the overall health of apple orchards. In conclusion, the field of ML for apple disease detection is a rapidly developing area with significant potential for improving the efficiency and effectiveness of disease management in apple orchards. Continued research and development in this area, along with the integration of ML models with precision agriculture techniques, will help to address the challenges and bring about significant advancements in this field.

REFERENCES

1. Arshleen Kaur, and Raman Chadha, "Classification Techniques for Disease Detection in Plants: A Systematic Review", *2023 International Conference on Artificial Intelligence and Smart Communication (AISC)*, pp. 1459–1463, 2023.
2. Vibhor Kumar Vishnoi, Krishan Kumar, Rajesh Kumar, Shashank Mohan, and Arfat Ahmad Khan, "Detection of Apple Plant Diseases Using Leaf Images through Convolutional Neural Network", *IEEE Access*, vol. 11, pp. 6594–6609, 2023.
3. Satish Kumar, Rakesh Kumar, and Meenu Gupta, "Analysis of Apple Plant Leaf Diseases Detection and Classification: A Review", *2022 Seventh International Conference on Parallel, Distributed and Grid Computing (PDGC)*, pp. 361–365, 2022.
4. Kamepalli Sujatha, Ketepalli Gayatri, M. Srikanth Yadav, N. Chandra Sekhara Rao, and Bandaru Srinivasa Rao, "Customized Deep CNN for Foliar Disease Prediction Based on Features Extracted from Apple Tree Leaves Images", *2022 International Interdisciplinary Humanitarian Conference for Sustainability (IIHC)*, pp. 193–197, 2022.
5. Harshit Singh, Kumud Saxena, and Ankit Kumar Jaiswal, "Apple Disease Classification Built on Deep Learning", *2022 3rd International Conference on Intelligent Engineering and Management (ICIEM)*, pp. 978–982, 2022.

6. Anupam Bonkra, Pramod Kumar Bhatt, Amandeep Kaur, and Sushil Kamboj, "Scientific Landscape and the Road Ahead for Deep Learning: Apple Leaves Disease Detection", *2023 International Conference on Artificial Intelligence and Smart Communication (AISC)*, pp. 869–873, 2023.

7. Astha Sharma, and Ashwini Kumar, "A Review: Classification and Detection of Plants Diseases Using Machine Learning and Soft Computing Techniques", *2022 4th International Conference on Artificial Intelligence and Speech Technology (AIST)*, pp. 1–6, 2022.

8. Kamepalli Sujatha, Ketepalli Gayatri, M. Srikanth Yadav, N. Chandra Sekhara Rao, and Bandaru Srinivasa Rao, "Customized Deep CNN for Foliar Disease Prediction Based on Features Extracted from Apple Tree Leaves Images", *2022 International Interdisciplinary Humanitarian Conference for Sustainability (IIHC)*, pp. 193–197, 2022.

9. Fatimah Al Heeti, and Muhammad Ilyas, "Performance Comparison of Convolutional Neural Network Models for Plant Leaf Disease Classification", *2022 International Symposium on Multidisciplinary Studies and Innovative Technologies (ISMSIT)*, pp. 386–391, 2022.

10. Dishan Sharma, Nitika Kapoor, and Parminder Singh, "Voting Classification Method with Clustering Method for the Plant Disease Detection", *2022 International Conference on Computational Intelligence and Sustainable Engineering Solutions (CISES)*, pp. 477–485, 2022.

11. S. Chakraborty, S. Paul, and M. Rahat-uz Zaman, "Prediction of Apple Leaf Diseases Using Multiclass Support Vector Machine," *2021 2nd International Conference on Robotics, Electrical and Signal Processing Techniques (ICREST)*, pp. 147–151, IEEE, 2021.

12. Swati Singh, et al. "Deep Learning Based Automated Detection of Diseases from Apple Leaf Images", *CMC-Computers, Materials & Continua*, vol. 71, no. 1, pp. 1849–1866, 2022.

13. L. Li, S. Zhang, and B. Wang, "Apple Leaf Disease Identification with a Small and Imbalanced Dataset Based on Lightweight Convolutional Networks", *Sensors*, vol. 22, no. 1, p. 173, 2022.

14. S. Baranwal, S. Khandelwal, and A. Arora, "Deep Learning Convolutional Neural Network for Apple Leaves Disease Detection", *Proceedings of International Conference on Sustainable Computing in Science, Technology and Management (SUSCOM), Amity University Rajasthan*, Jaipur-India, 2019.

15. M. K. Kaleem, et al., "A Modern Approach for Detection of Leaf Diseases Using Image Processing and ML Based SVM Classifier", *Turkish Journal of Computer and Mathematics Education (TURCOMAT)*, vol. 12, no. 13, pp. 3340–3347, 2021.

16. V. Srinidhi, A. Sahay, and K. Deeba, "Plant Pathology Disease Detection in Apple Leaves Using Deep Convolutional Neural Networks: Apple Leaves Disease Detection Using Efficientnet and Densenet," *2021 5th International Conference on Computing Methodologies and Communication (ICCMC)*, pp. 1119–1127, IEEE, 2021.

17. M. Shuaibu, W. S. Lee, J. Schueller, P. Gader, Y. K. Hong, and S. Kim, "Unsupervised Hyperspectral Band Selection for Apple Marssonina Blotch Detection", *Computers and Electronics in Agriculture*, vol. 148, pp. 45–53, 2018.

18. P. Bansal, R. Kumar, and S. Kumar, "Disease Detection in Apple Leaves Using Deep Convolutional Neural Network," *Agriculture*, vol. 11, no. 7, 2021. https://doi.org/10.3390/agriculture11070617

19. S. Manimegalai, and G. Sivakamasundari, "Apple Leaf Disease Identification Using Support Vector Machine", *Proceedings of the International Conference on Emerging Trends in Applications of Computing (ICETAC)*, pp. 1–4, 2017.

20. S. Alqethami, B. Almtanni, W. Alzhrani, and M. Alghamdi, "Disease Detection in Apple Leaves Using Image Processing Techniques," *Engineering, Technology & Applied Science Research*, vol. 12, no. 2, pp. 8335–8341, 2022.

21. K. P. Ferentinos, "Deep Learning Models for Plant Disease Detection and Diagnosis", *Computers and Electronics in Agriculture*, vol. 145, pp. 311–318, 2018.

22. S. P. Mohanty, D. P. Hughes, and M. Salathé, "Using Deep Learning for Image-Based Plant Disease Detection", *Frontiers in Plant Science*, vol. 7, 1419, 2016.

23. S. Sladojevic, M. Arsenovic, A. Anderla, D. Culibrk, and D. Stefanovic, "Deep Neural Networks Based Recognition of Plant Diseases by Leaf Image Classification", *Computational Intelligence and Neuroscience*, 2016. https://doi.org/10.1155/2016/3289801

24. G. Saradhambal, R. Dhivya, S. Latha, and R. Rajesh, "Plant Disease Detection and Its Solution Using Image Classification", *International Journal of Pure and Applied Mathematics*, vol. 119, pp. 879–883, Jan. 2018.

25. N. Ganatra, and A. Patel, "A Multiclass Plant Leaf Disease Detection Using Image Processing and Machine Learning Techniques", *International Journal on Emerging Technologies*, vol. 11, no. 2, pp. 1082–1086, 2020.

26. X. E. Pantazi, D. Moshou, and A. A. Tamouridou, "Automated Leaf Disease Detection in Different Crop Species through Image Features Analysis and One Class Classifiers", *Computers and Electronics in Agriculture*, vol. 156, pp. 96–104, Jan. 2019. https://doi.org/10.1016/j.compag.2018.11.005

27. V. K. Vishnoi, K. Kumar, and B. Kumar, "Plant Disease Detection Using Computational Intelligence and Image Processing", *Journal of Plant Diseases and Protection*, vol. 128, no. 1, pp. 19–53, Oct. 2021. https://doi.org/10.1007/s41348-020-00368-0

28. S. Ramesh et al., "Plant Disease Detection Using Machine Learning", *2018 International Conference on Design Innovations for 3Cs Compute Communicate Control (ICDI3C)*, pp. 41–45, Apr. 2018. https://doi.org/10.1109/ICDI3C.2018.00017

29. T. Fang, P. Chen, J. Zhang, and B. Wang, "Identification of Apple Leaf Diseases Based on Convolutional Neural Network", *Intelligent Computing Theories and Application*, pp. 553–564, 2019. https://doi.org/10.1007/978-3-030-26763-6_53

30. P. Jiang, Y. Chen, B. Liu, D. He, and C. Liang, "Real-Time Detection of Apple Leaf Diseases Using Deep Learning Approach Based on Improved Convolutional Neural Networks", *IEEE Access*, vol. 7, pp. 59069–59080, 2019. https://doi.org/10.1109/ACCESS.2019.2914929

31. B. Liu, Y. Zhang, D. He, and Y. Li, "Identification of Apple Leaf Diseases Based on Deep Convolutional Neural Networks", *Symmetry*, vol. 10, no. 1, Jan. 2018. https://doi.org/10.3390/sym10010011

32. F. Mohameth, C. Bingcai, and K. A. Sada, "Plant Disease Detection with Deep Learning and Feature Extraction Using Plant Village", *Journal of Computer and Communications*, vol. 8, no. 6, pp. 10–22, Jun. 2020. https://doi.org/10.4236/jcc.2020.86002

33. M. H. Saleem, J. Potgieter, and K. M. Arif, "Plant Disease Detection and Classification by Deep Learning", *Plants*, vol. 8, no. 11, Nov. 2019, Art. no. 468. https://doi.org/10.3390/plants8110468

34. P. Kulkarni, A. Karwande, T. Kolhe, S. Kamble, A. Joshi, and M. Wyawahare, "Plant Disease Detection Using Image Processing and Machine Learning," arXiv:2106.10698 [cs], Nov. 2021.

35. S. Singh, and S. Gupta, "Apple Scab and Marsonina Coronaria Diseases Detection in Apple Leaves Using Machine Learning", *International Journal of Pure and Applied Mathematics*, vol. 118, pp. 1151–1166, Jan. 2018.

36. T. Talaviya, D. Shah, N. Patel, H. Yagnik, and M. Shah, "Implementation of Artificial Intelligence in Agriculture for Optimisation of Irrigation and Application of Pesticides and Herbicides", *Artificial Intelligence in Agriculture*, vol. 4, pp. 58–73, 2020. https://doi.org/10.1016/j.aiia.2020.04.002

37. R. A. Bahn, A. A. K. Yehya, and R. Zurayk, "Digitalization for Sustainable Agri-Food Systems: Potential, Status, and Risks for the MENA Region", *Sustainability*, vol. 13, no. 6, 3223, 2021. https://doi.org/10.3390/su13063223

38. S. N. H. Bukhari, J. Webber, and A. Mehbodniya, "Decision Tree Based Ensemble Machine Learning Model for the Prediction of Zika Virus T-cell Epitopes as Potential Vaccine Candidates", *Scitific Reports*, vol. 12, 7810, 2022. https://doi.org/10.1038/s41598-022-11731-6

39. G. S. Raghavendra, S. Shyni Carmel Mary, Purnendu Bikash Acharjee, V. L. Varun, Syed Nisar Hussain Bukhari, Chiranjit Dutta, and Issah Abubakari Samori, "An Empirical Investigation in Analysing the Critical Factors of Artificial Intelligence in Influencing the Food Processing Industry: A Multivariate Analysis of Variance (MANOVA) Approach", *Journal of Food Quality*, vol. 2022, Article ID 2197717, 7 pages, 2022. https://doi.org/10.1155/2022/2197717

40. S. N. H. Bukhari, A. Jain, E. Haq, A. Mehbodniya, and J. Webber, "Ensemble Machine Learning Model to Predict SARS-CoV-2 T-Cell Epitopes as Potential Vaccine Targets", *Diagnostics*, vol. 11, no. 11, p. 1990, 2021. MDPI AG. https://doi.org/10.3390/diagnostics11111990

41. Sunil L. Bangare, Deepali Virmani, Girija Rani Karetla, Pankaj Chaudhary, Harveen Kaur, Syed Nisar Hussain Bukhari, and Shahajan Miah, "Forecasting the Applied Deep Learning Tools in Enhancing Food Quality for Heart Related Diseases Effectively: A Study Using Structural Equation Model Analysis", *Journal of Food Quality*, vol. 2022, Article ID 6987569, 8 pages, 2022. https://doi.org/10.1155/2022/6987569

42. C. M. Anoruo, S. N. H. Bukhari, and O. K. Nwofor, "Modeling and Spatial Characterization of Aerosols at Middle East AERONET Stations", *Theoretical and Applied Climatology*, vol. 152, pp. 617–625, 2023. https://doi.org/10.1007/s00704-023-04384-6

43. F. Masoodi, M. Quasim, S. Bukhari, S. Dixit, and S. Alam, *Applications of Machine Learning and Deep Learning on Biological Data*, CRC Press, 2023.

44. S. N. H. Bukhari, A. Jain, and E. Haq, "A Novel Ensemble Machine Learning Model for Prediction of Zika Virus T-Cell Epitopes", in Gupta, D., Polkowski, Z., Khanna, A., Bhattacharyya, S., and Castillo, O. (eds) *Proceedings of Data Analytics and Management. Lecture Notes on Data Engineering and Communications Technologies*, vol. 91. Springer, Singapore, 2022. https://doi.org/10.1007/978-981-16-6285-0_23

45. S. N. H. Bukhari, F. Masoodi, M. A. Dar, N. I. Wani, A. Sajad, and G. Hussain, "Prediction of Erythemato-Squamous Diseases Using Machine Learning," in *Auerbach Publications eBooks*, 2023, pp. 87–96. http://doi.org/10.1201/9781003328780-6

46. S. N. H. Bukhari, A. Jain, E. Haq, A. Mehbodniya, and J. Webber, "Machine Learning Techniques for the Prediction of B-Cell and T-Cell Epitopes as Potential Vaccine Targets with a Specific Focus on SARS-CoV-2 Pathogen: A Review", *Pathogens*, vol. 11, no. 2, p. 146, 2022. MDPIAG. http://doi.org/10.3390/pathogens11020146

47. S. Nisar, H. Bukhari, and M. A. Dar, "Using Random Forest to Predict T-Cell Epitopes of Dengue Virus," *Advances and Applications in Mathematical Sciences*, vol. 20, no. 11, pp. 2543–2547, 2021.

3

Intelligent Inputs Revolutionizing Agriculture: An Analytical Study

Nelofar Ara, Sani Mustapha Kura, VK Aswathy,
and Mohammad Amin Wani
Lovely Professional University, Phagwara, India

3.1 INTRODUCTION

Agriculture plays a central role in the economy of every country. The need for food is increasing every day, along with the world population [1]. Currently, the traditional techniques of farmers cannot meet the demand. Various innovative methods of automation have been created to meet these needs, offering great career opportunities to many people in this field. Throughout human history, technology has long been used in agriculture to improve efficiency and reduce the labor intensity of agriculture. Since the invention of farming, humans and agriculture have evolved from better plows to irrigation, tractors, and modern artificial intelligence (AI) [2]. The increasing and more accessible availability of computer vision could be a major advance in this field. As significant global changes occur in our climate, ecology, and food demand, AI could transform 21st-century agriculture in the following ways:

- Improving time, labor, and resource efficiency
- Increasing the sustainability of the environment
- Smatter allocation of resources
- Providing real-time monitoring to encourage improved produce quality and health.

DOI: 10.1201/9781003485179-3

For AI to achieve this, the agricultural sector must change. This calls for more technical and pedagogical investment in the agricultural sector to transform farmers' knowledge of their "field" into AI training. However, agricultural ingenuity and adaptation are zero innovative. For farmers, computer vision and agricultural robotics are the latest ways to use new technology to meet global food needs and improve food security. There is no doubt that today's work practices, agricultural yields, and quality exceed those of 500 or even 50 years ago. However, there is still a lot of room for improvement. By 2050, the planet will be home to 9.9 billion people, after which food demand is expected to increase from 35% to 56%. Not to mention how climate change is depleting natural resources, such as water and agricultural land. Fortunately, technology offers us another resource: AI. The agricultural sector is evolving at a whole new pace as advances in disease detection, predictive analytics, and computer vision technologies are used to monitor crops and soil. In addition to the potential, there is also rapidly growing interest and investment: Global "smart" agriculture, which includes machine learning and AI, is projected to triple to $15.3 billion by 2025, according to *Forbes*. Researchers predict that the market for AI in agriculture will grow with the combination. Compound annual growth rate (CAGR) of 20% to reach $2.5 billion by 2026. And that's just the beginning! This chapter examines the most innovative applications of AI that are reshaping the agricultural industry, including the following.

3.1.1 Crop and Soil Monitoring

The micro- and macronutrient composition of the soil directly affects the quantity, quality, and general health of the crop [3]. For farmers to maximize production efficiency, it is important to control the stages of the plants' development after planting them in the ground. Understanding the relationship between crop growth and the environment is critical to making changes that improve crop health. Now historically, human observations and assessments have been used to assess soil and plant health. However, this approach is neither accurate nor up-to-date. Instead, drones (unmanned aerial vehicles (UAVs)) are now used to capture aerial imagery and train computer vision models that can be used to intelligently monitor crop and soil conditions. Farmers can act quickly by learning specific problem areas using AI models.

3.1.2 Observing Crop Maturity

Such a labor-intensive task, which involves manually monitoring the growth phase of a wheat head, can be assisted by AI in the accuracy of agriculture. The researchers were able to build a "two-stage coarse to fine wheat ear detection mechanism" by collecting photographs of wheat at different stages of "orientation" over three years and under different lighting conditions. Farmers no longer had to go to the field every day to check their crops because this computer vision model identified the growth stages of the wheat more accurately than human observation. Another study looked at the accuracy of tomato ripeness detection using computer vision. The researchers developed an algorithm that tests the color of five different tomato ingredients and uses the results to determine the degree of ripeness of the tomato. The method attained a successful discovery and classification rate of 99.31%. Excessive care and overestimation of the development and maturity of crops require a lot of work from farmers. However, AI shows its ability to handle a significant part of the work easily and perfectly.

3.1.3 Hitting the Ground with Computer Vision

Returning to the significance of soil, a separate study examined the accuracy of computer vision images of soil structure and soil organic matter (SOM). Generally, farmers have to collect soil samples from the soil and take them to the laboratory for laborious and time-consuming evaluation [4]. Instead, the researchers decided to see if they could train an algorithm to perform the same task using image data from an inexpensive handheld microscope. The computer vision model was able to provide estimates of SOM and sand content that were as accurate as expensive laboratory processing. As a result, computer vision can not only eliminate a significant portion of the labor-intensive manual labor required in crop and soil monitoring but often does so more efficiently than humans.

3.1.4 Insect and Plant Disease Detection

We've seen how computer images powered by AI can identify and assess crop maturity and soil quality, but what about less predictable farming situations? Using image recognition technology based on deep learning, it is now possible to automatically detect plant diseases and pests. It works by creating models that can monitor plant health using image classification, recognition, and segmentation techniques.

3.1.5 Keeping Out the Bad Apples (Diagnosing Disease Severity)

A study on apple black rot – something we all definitely don't want on our apples! – gives a good example of this in practice. The Deep Convolutional Neural Network was trained using photographs of apple rot, which botanists classified into four levels of severity. The computer vision alternative, as in our previous cases, requires labor-intensive human search and evaluation. Luckily for agriculturalists, the AI model used in this study is 90.4% accurate in detecting and diagnosing the severity of the disease! In an additional study, the researchers went even further by using the advanced YOLO v3 algorithm to detect many diseases and pests on tomato plants. Using a digital camera and smartphone, the researchers photographed the presence of 12 different diseases or pests in nearby tomato greenhouses. The model was trained with images of different resolutions and object sizes, after which it was able to identify diseases and pests with 92.39% accuracy in less than 20.39 milliseconds.

3.1.6 Finding Bugs with Code

The analysts, to begin with, set up a sticky trap to capture six diverse sorts of flying creepy crawlies and take real-time photographs. They used YOLO object recognition for object detection and coarse computations and support vector machines (SVMs) using global features for object classification and fine computations. When all was said and done, their computer vision show was able to number them with 92.5% exactness and recognize between bees, flies, mosquitoes, moths, wasps, and natural product flies at 90.18%. Computer vision-based AI has the potential to monitor the health of our food systems, according to this study. In addition, it reduces labor inefficiencies while maintaining the accuracy of observations.

3.1.7 Livestock Health Monitoring

So far, we've mainly talked about crops, but farming includes more than wheat, tomatoes, and apples. Our agricultural system also relies heavily on animals, which require a little more management than plants. Can computer vision keep up with pigs, cows, and chickens? Cattle-Eye is a great example of agricultural AI. They use aerial cameras and computer vision algorithms to monitor livestock health and behavior. This means that the farmer does not necessarily have to stand next to the cow to see the problem. By remotely monitoring and tracking flocks in real time, farmers can be alerted as soon as problems are detected. Livestock aren't the only victims. Additionally, computer vision enables the following:

- Count animals, look for sickness, spot odd conduct, and keep an eye on significant activities, such as giving birth.
- Gather information using drones and cameras.
- Combine this with other technologies to educate farmers on animal health and food, as well as water availability.

3.1.8 Intelligent Spraying

UAVs equipped with computer vision. AI enables automatic and consistent application of pesticides or fertilizers across the field. Real-time detection of the target spray area enables UAV sprayers to work with high precision in terms of spray area and spray volume. As a result, there is significantly less chance of contaminating crops, livestock, people, and water systems. Although this place has a lot of potential, there are still challenges. For example, using multiple UAVs to spray a large area is much more efficient, but it can be difficult to determine specific mission cycles and flight paths for each vessel. However, this does not always mean intelligent injection is no longer an option [5]. Virginia Tech researchers have developed a smart spraying system based on servo-driven sprayers that use computer vision to identify weeds. A camera attached to the sprayer records the geographic location of the weed and assesses the size, shape, and color of each elusive plant to deliver the exact amount of herbicide to the target. In other words, it acts as a kind of herbicide. Unlike the Terminator, the computer vision system can shoot with such precision that it prevents additional damage to prey or the environment.

3.1.9 Automatic Weeding

The use of AI extends beyond weed control sprinklers. Other, more direct computer vision robots remove unwanted plants. While computer vision can spot an insect or a chicken behaving strangely, weed detection doesn't save the farmer a lot of work. For AI to be even more useful, it needs to find and remove weeds. Being able to remove weeds by hand not only saves a farmer's work but also reduces the need for pesticides, making the entire farming process much more environmentally friendly and sustainable.

3.1.10 Robots in the Weeds

With target detection, weeds and crops can be easily distinguished from each other. But the real power is seen when computer vision and machine-learning techniques are applied to the construction of weeding robots. All of this is a great introduction to BoniRob, an agricultural robot that uses cameras and image recognition technologies to look for weeds and then chase them out of the ground. It acquires the ability to differentiate between weeds and crops through visual training based on leaf size, shape, and color. In this way, BoniRob can remove unwanted plants from the field without having to worry about destroying anything irreplaceable. To accomplish this, researchers are creating agricultural robots for the detection of weeds and soil moisture. It can move around the field in this way, removing plants and drenching the soil with just the right amount of water. According to the test results, the plant classification and eradication rate of this technology is 90% or more, while the moisture content of the deep soil is maintained at 80%–10%.

3.1.11 Aerial Survey and Imaging

It's no surprise that computer vision has great applications for mapping land and monitoring crops and livestock. However, this does not diminish its importance for smart agriculture. AI can analyze satellite and drone footage to help farmers monitor their livestock and crops. They would be notified as soon as something unusual appeared, so they wouldn't have to constantly check the fields in person. Aerial photography improves the accuracy and efficiency of pesticide application. As mentioned earlier, ensuring that pesticides reach their intended destination saves costs and benefits the environment.

3.1.12 Produce Grading and Sorting

AI-based computer vision can continue to support farmers after harvest. With the help of imaging algorithms, defects, diseases, and pests can be detected during plant growth, and "good" products can be distinguished from defective or unpleasant products. Computer vision can automate sorting and sorting by analyzing the size, shape, color and quantity of fruits and vegetables with much greater accuracy and speed than qualified professionals.

3.1.13 Picture Perfect Produce

To distinguish between carrots with surface defects or carrots that are not the right size and shape, the researchers created an automatic method that sorts them. A "good" carrot must, therefore, have the correct shape ("convex polygon"). and must not have fibrous roots or surface cracks. Smart injection is no longer an option [5]. Virginia Tech researchers have developed a smart spraying system based on servo-driven sprayers that use computer vision to identify weeds. A camera attached to the sprayer records the geographic location of the weed and assesses the size, shape, and color of each elusive plant to deliver the exact amount of herbicide to the target. In other words, it acts as a kind of herbicide. In both cases, a huge amount of labor-intensive physical labor was saved. Figure 3.1 shows this clearly:

On the left side:

- Image of a field with weeds
- Camera attached to a servo-driven sprayer
- Arrows indicating data capture (location, size, shape, color)
- Herbicide spraying action

On the right side:

- Agricultural robots (depicted as robots with various tools)
- Crop soil monitoring (depicted as soil with sensors)
- Disease diagnosis (depicted as a plant with a magnifying glass)

Virginia Tech's smart spraying system utilizes computer vision for weed detection, optimizing herbicide use and reducing labor. The innovation integrates with other agricultural technologies for comprehensive farm management.

FIGURE 3.1
Smart spraying system by Virginia Tech-Greenhouse controlled environment agriculture.

Source: https://www.bing.com/images/search/yield-gruetzmacher-ai-automation-greenhouse-controlled-environment-agriculture

3.2 ANALYZING FARM DATA USING AI

It's amazing how farmers can now collect and analyze data from their fields due to AI; weather, temperature, water consumption, soil conditions, etc., can also be recorded in real time. Every day, thousands of field data points are recorded in the field. AI technology is already helping farmers improve yields through precision agriculture, which improves crop selection, hybrid seed selection, resource use, crop quality, and crop precision [6]. Precision agriculture uses AI technology to detect diseases, pests, and malnutrition in fields. AI sensors can detect weeds, target them, and choose which pesticides and herbicides to apply within a defined buffer zone [7]. This helps farmers maximize the number of pesticides and herbicides they use. In addition, AI helps farmers create seasonal forecasting models to increase crop production and accuracy. Using technology to predict weather trends in the coming months can also help farmers make smarter decisions. Small farms in less developed countries are the main target for seasonal forecasts, as information and data can be scarce. These small farmers must produce the best crops because they produce 70% of the world's crops. Farmers use AI-powered drone-based cameras in addition to topographic data to better photograph their produce and analyze site photos in real time to identify potential problems and areas for improvement. Drones monitor the product more efficiently and can cover a larger area than humans. The use of AI-based image recognition techniques for disease diagnosis, insect infestation detection, and plant identification is also becoming more widespread. A program based on X-Fito understanding was created by Cruz et al. (2017) to identify signs of Olive Quick Decline Syndrome (OQDS) [8]. The system uses transform-focused neural networks (DP-CNN) and new abstraction-level fusion techniques to improve diagnostic accuracy. Crus's PCA (Principal Component Analysis) system for industrial applications is used in a similar manner by farmers to make quick management decisions and reduce disease (Ibid).

3.2.1 Yield Management Using AI

AI, cloud technology, satellite imagery, and advanced analytics have created a modern smart farming ecosystem. By combining these technologies, farmers can increase average yields and better control prices. Cortana Intelligence Suite and machine learning are two advisory services Microsoft is now offering to farmers in Andhra Pradesh. This pilot project used AI applications in agriculture to provide data, soil preparation, soil test-based fertilization, seed treatment, ideal sowing depth, and more. Mobile robots and field sensors support digital farming robots, multifunction cameras, and laser scanners in blind spots and wireless areas.

3.2.1.1 *Tackling the Labor Challenge*

Due to the massive migration of labor to the city, farm operations are currently stalled. AI-powered robots improve employee productivity in a number of different ways. These robots can collect speedier and discover and expel weeds more precisely, lessening working costs and labor. Ranchers are, as of now, inquiring about chatbots to offer assistance [9]. Chatbots bolster agriculturists by replying to their questions and giving counsel and direction on particular agrarian issues.

3.2.1.1.1 Applications of AI in Agriculture

In the above image, AI technologies are depicted on the left side, showcasing their role in various aspects of agriculture, including crop yield production and price forecast, intelligent spraying, and providing productive insights. On the right side, AI is shown to facilitate connections with agriculture robots, crop soil monitoring, and disease diagnosis. This illustrates the comprehensive impact of AI in optimizing agricultural processes, from enhancing productivity to enabling precision management practices and proactive disease control.

As with traditional farming methods, farmers face many challenges. In this field, AI is often used to solve these problems (Figure 3.2). AI is a

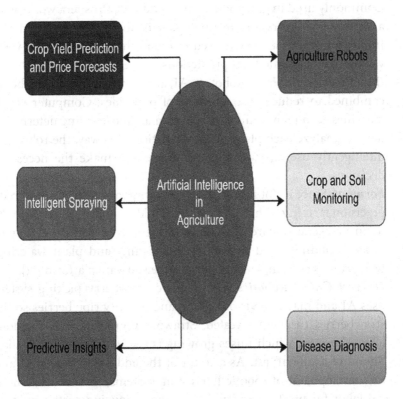

FIGURE 3.2
Applications of AI in agriculture.

Source: Prepared by the researchers

pioneering technology in agriculture today. This has benefits in producing healthier crops, pest control, soil monitoring, and more [10].

3.3 AI START-UPS IN AGRICULTURE

Based on data inputs, attempts are being made worldwide to automate the operations of agricultural businesses and use AI and related technologies to streamline agriculture [8]. This has led to the creation of countless start-ups.

- *Prospera*: The start-up has developed a cloud-based application that integrates all the data available to farmers, including water and soil sensor data, aerial photography, and more. It then connects to field devices that analyze the data and recommend desired results. Commonly used in greenhouses and outdoors, Prospera works with a variety of sensors and technologies, including computer vision. Input from these sensors is used to establish relationships between various data tags and their predictions.
- *Blue River Technology*: Robotics, AI, and computer vision are being combined to reduce costs and use of pesticides. Computer vision identifies each plant individually, and machine learning determines how to analyze each plant's characteristics. This way, the robot can intelligently use agricultural machinery and make the necessary decisions.
- *Formbot*: Since its founding in 2011, this organization has promoted precision farming techniques for smallholders and engaged the public in precision research. It allows farmers to perform a variety of tasks, including weed detection, soil testing, and plant watering, using robots running on an open-source software platform [11].
- *Harvest CROO Robotics*: This robotic strawberry picking system uses AI and machine vision to find and identify ripe berries so the strawberries can be harvested. Strawberry growers face a serious labor shortage, which raises growing costs and increases the likelihood of a subharvest. As a result of the additional use of AI and the development of robotic harvesting systems, the need for manual labor for producers would be reduced, logging costs would be reduced, and overall competitiveness would increase.

- *Gramophone (Agstack Technologies)*: Using image recognition skills, the right information, methods, and ingredients are available at the right time so farmers get the best possible harvest. The company predicts pests and diseases, predicts when food prices will increase production, and recommends products to farmers using AI and machine learning.
- *Jivabhumi*: Create an "intelligent" agricultural market to balance the often insufficient supply and demand of agricultural products [12]. A revolutionary food aggregation system that combines agricultural products, electronic marketing services, and innovation. Technologies such as blockchain are used to collect information about products at various points in the supply chain.

3.4 CHALLENGES IN THE ADOPTION OF AI

Despite the enormous promise of AI in agriculture, there are still many farms that do not have access to cutting-edge machine-learning techniques. The vulnerability of agriculture to external forces such as weather, soil quality, and pests is also crucial. Due to changes in external circumstances, a decision that seemed wise in the planning phase did not necessarily turn out to be the best. AI systems require large amounts of data to train algorithms to make accurate predictions. Finding temporal data for large agricultural areas is difficult, while collecting geographic data is easy. As the data infrastructure evolves, building a robust machine-learning model takes time. This is one reason why AI is not used in field solutions but in agricultural products, such as seeds, fertilizers, and pesticides. When it comes to farmers who are in control of their decisions and are able to make their own decisions and predictions depending on the circumstances, agriculture is still at an extremely early stage.

3.5 CONCLUSION

Summarizing the aforementioned research/study, we conclude that in all industries, including education, banking, robotics, agriculture, etc., the development of AI is considered one of the most important technologies.

This is changing the agricultural sector and has a major impact on it. AI protects the agricultural sector from a number of challenges, such as food security, population growth, climate change, and labor problems [13]. Modern agriculture and AI have both risen to unprecedented heights. AI has improved crop production, real-time monitoring, harvesting, processing, and marketing. A variety of high-tech computer systems are developed to detect several key criteria, including weed detection, crop detection, crop quality, and many others [14]. For applications to progress, they must be flexible to explore the many applications of AI in agriculture. It must be able to adapt to changing environments, support real-time decision-making, and obtain relevant information using appropriate platforms/structures. Another major drawback is that the cost of many options in the agricultural sector is very high. For the technology to be available at the farm level, the solutions must be more cost-effective. An open-source platform produces a more organized solution that allows farmers to deploy faster and better. The development of computer vision-based AI techniques requires a learning (preparing) approach that requires the collection and examination of a heap of illustrations in normal and energetic situations that precisely duplicate on-site conditions. AI technology usually advances by generating higher-quality data, allowing processors to deal with many imaging problems, such as inadequate lighting, inappropriate cropping, and misaligned objects [15]. Mobile devices can be combined with these algorithms and AI technology to provide an effective platform for pest and disease detection and pesticide mapping. The use and costs of pesticides can be significantly reduced, thus reducing their environmental impact, by using agrochemicals in the right amount and using pesticides only when necessary.

ACKNOWLEDGMENT

This work has been written by reviewing earlier studies (published/unpublished) by the corresponding author (Nelofar Ara) and supporting coauthors (Sani Mustapha Kura, Dr. Aswathy VK, and Dr. Mohmmad Amin Wani.

REFERENCES

1. Parr, C. S., Lemay, D. G., Owen, C. L., Woodward Greene, M. J., & Sun, J. (n.d.). Multi-modal AI to improve agriculture. *IT Professional*, 23, 53–57.

2. Kant, A., Ajay, S., Vivek, G., Jayesh, R., & Jeremy, J. (March 2021). *Artificial intelligence for agriculture innovation.* Retrieved from https://www3.weforum.org/docs/WEF_Artificial_Intelligence_for_Agriculture_Innovation_2021

3. Pandey, N., & Kamboj, N. (2022). Consequences of soil quality on crop yield: An update. *Indian Journal of Ecology,* 49(4), 1406–1416.

4. Islam, M. R., Singh, B., & Dijkstra, F. A. (2022). Stabilisation of soil organic matter: Interactions between clay and microbes. *Biogeochemistry,* 160(2), 145–158.

5. Alex, N., Sobin, C. C., & Ali, J. (2023). A comprehensive study on smart agriculture applications in India. *Wireless Personal Communications,* 129(4), 2345–2385.

6. Javaid, M., Haleem, A., Khan, I. H., & Suman, R. (2023). Understanding the potential applications of artificial intelligence in agriculture sector. *Advanced Agrochem,* 2(1), 15–30.

7. Talaviya, T., Shah, D., Patel, N., Yagnik, H., & Shah, M. (2020). Implementation of artificial intelligence in agriculture for optimization of irrigation and application of pesticides and herbicides. *Artificial Intelligence in Agriculture,* 4, 58–73.

8. Cruz, J., Flórez, J., Torres, R., Urquiza, M., Gutiérrez, J. A., Guzmán, F., & Ortiz, C. C. (2017). Antimicrobial activity of a new synthetic peptide loaded in polylactic acid or poly (lactic-co-glycolic) acid nanoparticles against Pseudomonas aeruginosa, Escherichia coli O157: H7 and methicillin resistant Staphylococcus aureus (MRSA). *Nanotechnology,* 28(13), 135102.

9. Deng, L., Du, H., & Han, Z. (2017). A carrot sorting system using machine vision technique. *Applied Engineering in Agriculture,* 33(2), 149–156.

10. Ramos-Giraldo, P., Reberg-Horton, C., Locke, A. M., Mirsky, S., & Lobaton, E. (May/Jun. 2020). Drought stress detection using low-cost computer vision system and machine learning techniques. *IT Professional,* 22(3), 27–29.

11. Subeesh, A., & Mehta, C. R. (2021). Automation and digitization of agriculture using artificial intelligence and Internet of Things. *Artificial Intelligence in Agriculture,* 5, 278–291.

12. Maheswari, R., Ashok, K. R., & Prahadeeswaran, M. (2008). Precision farming technology, adoption decisions and productivity of vegetables in resource-poor environments. *Agricultural Economics Research Review,* 21, 415–424.

13. Revathi, A., & Poonguzhali, S. (2023). The role of AIoT-based automation systems using UAVs in smart agriculture. In Divya Mishra, & Shanu Sharma (Eds.), *Revolutionizing industrial automation through the convergence of artificial intelligence and the Internet of Things* (pp. 100–117). IGI Global.

14. Hatfield, J. L., Cryder, M., & Basso, B. (May/Jun. 2020). Remote sensing: Advancing the science and the applications to transform agriculture. *IT Professional,* 22(3), 42–45.

15. Peters, D. P. C., Savoy, H. M., Ramirez, G. A., & Huang, H. H. (May/Jun. 2020). AI with ML recommender systems for agricultural research. *IT Professional,* 22(3), 30–32.

4

Case Studies on the Initiatives and Success Stories of Edge AI Systems for Agriculture

Thamizhiniyan Natarajan and Shanmugavadivu Pichai
Gandhigram Rural Institute, Chinnalapatti, India

4.1 INTRODUCTION

An era of unprecedented transformation is being ushered into the agricultural sector, driven by the fusion of cutting-edge technology with age-old farming practices, spearheaded by the advent of Edge AI (Artificial Intelligence at the Edge). Within this dynamic landscape, Edge AI is positioned as a catalyst for revolutionizing agriculture, offering a multitude of solutions to address long-standing challenges and usher in a new era of smart farming.

For millennia, agriculture has served as the bedrock of human survival, providing sustenance, livelihoods, and economic prosperity to nations across the globe. However, the demands on this critical sector have never been more significant. With the world's population surging and climate change disrupting traditional weather patterns, agriculture faces mounting pressure to meet the growing need for food production while simultaneously contending with limited resources and environmental degradation.

The role of Edge AI in agriculture becomes apparent as the convergence of artificial intelligence and the Internet of Things (IoT) empowers farms with real-time insights, facilitating immediate responses to changing conditions and enabling data-driven precision agriculture.

DOI: 10.1201/9781003485179-4

The pivotal role of Edge AI in agriculture is highlighted through a comprehensive exploration of a curated collection of case studies drawn from diverse research papers. These case studies encapsulate the transformative potential of Edge AI, showcasing its ability to reshape the farming landscape in multifaceted ways.

From the utilization of AI-powered cameras for detecting plant diseases in the field to the remote monitoring of livestock health through edge devices, from precision irrigation systems that fine-tune water distribution based on real-time data to autonomous farming solutions that minimize manual labor while maximizing productivity, a spectrum of agricultural applications is covered. The functional features and benefits of Edge AI systems are illuminated, and the socio-economic impact, adoption, and implications for farming communities and the environment are delved into.

As this journey through the realm of Edge AI in agriculture is embarked upon, it becomes evident that more than a technological evolution is being witnessed. A fundamental shift is underway, one that promises to redefine the very nature of farming itself. Empowering farmers with tools for informed decision-making, optimizing resource allocation, minimizing environmental footprints, and holding the key to global food security in a world marked by uncertainties are the promises of Edge AI.

In each case study, stories of innovation, resilience, and transformation are unveiled. These narratives underscore the profound influence of Edge AI on agriculture, reinforcing its status as a pivotal force in shaping the future of food production, promoting sustainability, fostering innovation, and ensuring food security in an increasingly complex and interconnected world. As the details of each case study are delved into, an invitation is extended to witness first-hand how Edge AI is turning the age-old practice of farming into a cutting-edge science.

The chapter's structure unfolds as follows: Section 4.2 explores related works on Edge AI in agriculture, offering insights into the field's existing research. In Section 4.3, an extensive examination of case studies is presented, offering a deeper understanding of practical applications. Section 4.4 comprehensively addresses the utilization, benefits, socio-economic impact, and adoption of Edge AI in agriculture. Lastly, Section 4.5 serves as the conclusion, summarizing the chapter's key findings and insights.

4.2 RELATED WORK

In a world driven by data and constant innovation, the combination of artificial intelligence (AI) with the untapped potential of edge computing and the IoT is recognized as a revolutionary force ready to transform industries, societies, and our daily lives. AI, known for its ability to uncover patterns, predict outcomes, and automate tasks, has become an integral part of modern existence. From recommendation algorithms that guide our entertainment choices to autonomous vehicles navigating busy streets, AI silently orchestrates the marvels of our time.

However, as AI's influence proliferated, an inherent challenge materialized. Traditional AI systems, predominantly housed in remote data centers and cloud infrastructures, dazzled with their prowess but were hampered by latency issues. The delay between data generation and its subsequent analysis, while insignificant for specific use cases, emerged as a critical factor in situations requiring immediate decision-making. This is where the concept of "edge" becomes relevant [1, 2].

4.2.1 The Edge AI Advantage

Edge AI signifies a radical departure from the centralized AI paradigm. It positions computation, data storage, and processing at the very periphery of the network – right where the action unfolds. By doing so, it collapses the time lag between data generation and analysis, ushering in a new era of immediacy and responsiveness [1]. The advantages of Edge AI are as follows:

- *Real-Time Decision-Making*: One of the foremost advantages of Edge AI is its capacity for real-time decision-making. Devices equipped with Edge AI can process data instantaneously, enabling split-second responses. This capability is a game-changer for applications like autonomous vehicles, industrial automation, and health-care monitoring, where even the briefest delay can have significant consequences [3].
- *Enhanced Privacy and Security*: Edge AI also bolsters data privacy and security. Since data is processed locally on the device, sensitive information need not traverse networks, reducing the risk of exposure. This is particularly vital in applications like health care, where patient data must be handled with utmost confidentiality [4].

- *Bandwidth Efficiency*: Edge AI optimizes bandwidth usage by processing data locally. This reduces the strain on network infrastructure, making it suitable for IoT deployments in remote or bandwidth-constrained environments [5].
- *Scalability*: Edge AI exhibits remarkable scalability, making it suitable for an extensive array of devices and usage scenarios. It can be customized to align with the distinct requirements of various applications, spanning from small sensors to robust industrial machinery [6].

4.2.2 The Evolution of Edge AI

The concept of Edge AI didn't emerge overnight but evolved over the years as a response to the growing demands of real-time, distributed computing.

Historically, AI relied on centralized computing resources, with data sent to remote servers for processing. However, as AI applications expanded into areas like autonomous vehicles, industrial automation, and IoT, the limitations of this approach became evident. The latency introduced by round-trip data transfer to distant data centers was unacceptable in scenarios where immediate decisions were crucial.

The concept of "edge computing" emerged to define a novel approach where computing resources were positioned nearer to the data source within the network. This transformation laid the foundation for the inception of Edge AI [1]. Leveraging advancements in hardware capabilities, Edge AI devices gained the capacity to execute intricate AI computations locally.

In agriculture, for instance, Edge AI–enabled tractors and drones to analyze field data in real-time, optimizing planting and harvesting decisions. In health care, wearable devices equipped with Edge AI became capable of monitoring vital signs and alerting users to potential health issues without requiring constant cloud connectivity.

4.2.3 Edge AI and IoT in Agriculture

Edge AI's impact extends into the agricultural sector, ushering in a new era of precision farming and sustainable practices.

- *Precision Agriculture*: Edge AI empowers agricultural machinery with the capability of making immediate decisions regarding planting, harvesting, and irrigation. Through sensors and cameras installed on tractors and drones, data related to soil conditions, crop health, and weather patterns is gathered. This information is promptly processed by Edge AI, allowing farmers to enhance their operations, minimize resource utilization, and boost crop yields [7].

- *Crop Monitoring*: Edge AI–powered cameras placed throughout fields can monitor crop health, identifying signs of disease, pest infestations, or nutrient deficiencies. This data can be used to trigger automated responses, such as targeted pesticide application, thereby reducing the need for broad-spectrum treatments and minimizing environmental impact [8].
- *Livestock Management*: In livestock farming, Edge AI can monitor the health and behavior of animals in real time. Wearable sensors and cameras can detect signs of distress or illness, allowing for early intervention. Additionally, AI-powered feeders can dispense precise amounts of feed, reducing waste and optimizing nutrition [9].

4.2.4 Challenges and Considerations

While Edge AI offers tremendous potential in agriculture, it comes with its own set of challenges:

- *Connectivity*: In remote agricultural areas, connectivity can be limited. Reliable network infrastructure is essential to ensure seamless data transfer between edge devices and central systems [1].
- *Data Privacy*: Edge AI collects a vast amount of data, including information about crop conditions and farm operations. Ensuring the privacy and security of this data is paramount [10].
- *Scalability*: The agricultural sector encompasses a wide range of operations, from small family farms to large commercial enterprises. Edge AI solutions must be scalable and adaptable to meet the needs of diverse agricultural settings [10].

In the world of Edge AI, there are endless opportunities for innovation. It's a space where intelligence operates right at the edge of what's possible, where data quickly turns into useful information, and where Edge AI's history and future are closely intertwined.

4.3 TRANSFORMATIVE TALES: CASE STUDIES SHOWCASING EDGE AI IN AGRICULTURE

In this section, the focus shifts toward real-world applications of Edge AI within the agricultural sector. A series of case studies is delved into, demonstrating how Edge AI systems have been effectively put into operation to tackle pressing challenges and promote innovation in farming practices. Both national and international contexts are covered by these case studies, underlining the versatility and transformative potential of Edge AI in agriculture. In each instance, attention is drawn to the unique approaches taken to leverage Edge AI technologies for the optimization of operations, enhancement of crop yields, improved resource management, and, ultimately, the contribution to a more sustainable and productive agricultural landscape. These case studies stand as concrete examples of the capability of Edge AI to revolutionize farming, thereby paving the way for a future in agriculture that is smarter, more environmentally friendly, and highly efficient.

4.3.1 Enhancing Plant Disease Classification with Edge AI on IoT Devices

In the pursuit of more efficient and sustainable farming practices, Trong-Minh Hoang and their team [11] have introduced a ground-breaking approach to plant disease classification using deep convolutional neural networks (DCNN) on IoT edge devices. Traditional methods for disease detection in agriculture have often been limited in their scope and efficiency. To address these limitations, the researchers have developed a lightweight and highly efficient DCNN model that is well-suited for deployment on resource-constrained IoT devices. This case study underscores the pivotal role of edge computing and AI solutions in the realm of precision agriculture, shedding light on how machine-learning algorithms can significantly enhance the timely and accurate detection of diseases in smart farming.

4.3.1.1 Challenges Addressed

- *Resource Constraints*: Edge devices, typically employed in remote farming areas, often have limited computational resources. This constraint hampers the utilization of complex machine-learning models, which are essential for accurate disease identification.

- *Model Size*: Downsizing machine-learning models to fit edge devices without sacrificing accuracy is a complex problem. Shrinking models too aggressively can lead to a loss of critical information and a decrease in identification accuracy.
- *Real-World Applicability*: Models must perform well not only in laboratory conditions but also in real-world agricultural settings, where environmental factors can introduce significant variability.

4.3.1.2 Key Benefits

- *Real-Time Insights*: Edge AI enables farmers to receive real-time updates on the health of their crops, allowing for immediate intervention when disease outbreaks occur.
- *Proactive Disease Management*: With timely detection, farmers can take proactive measures to prevent the spread of diseases, minimizing crop losses.
- *Improved Yields*: By addressing diseases promptly, farmers can optimize their agricultural practices and enhance overall crop yields.
- *Resource Efficiency*: The efficiency of the DCNN model ensures that it can operate effectively on resource-constrained IoT devices, making it a practical solution for real-world applications in precision agriculture.

4.3.1.3 Results and Implications

The DCNN model developed by Trong-Minh Hoang and their team has demonstrated remarkable accuracy in plant disease classification, outperforming other machine-learning models while maintaining high efficiency. This case study serves as a testament to the potential of machine-learning algorithms in achieving timely and accurate disease detection on IoT edge devices. Furthermore, it highlights the transformative impact of Edge AI in revolutionizing the agriculture industry and bolstering global food security.

4.3.2 Securing Smart Climate Agriculture with Edge AI and Blockchain Integration

In the modern era, technology plays a pivotal role in advancing various sectors, including agriculture. To address the unique challenges faced by agriculture, especially in countries like India, there is a growing need for technological modernization. This case study, inspired by the work of Li

Ting et al. [12], explores the development of an intelligent agricultural system that leverages cutting-edge technology to monitor and optimize essential climate and environmental parameters. By integrating IoT and blockchain technology, this system offers efficient and secure decision-making capabilities, ultimately enhancing plant health and agricultural productivity.

4.3.2.1 Challenges Addressed

- *Data Privacy and Security*: With the increasing number of IoT devices connecting to agricultural systems, the case study prioritizes data privacy and security. It accomplishes this by implementing blockchain technology to guarantee the secure transmission and storage of sensitive data.
- *Real-Time Monitoring*: Successful agriculture requires continuous monitoring of crucial parameters, including temperature, humidity, soil moisture, and light intensity. The proposed system integrates IoT sensors to collect and transmit this data in real time, enabling prompt analysis and decision-making.
- *Water Management*: Water conservation is essential in agriculture. The system employs intelligent fuzzy logic to make well-informed decisions regarding the timing and quantity of irrigation, ensuring efficient water usage and optimal plant health.

4.3.2.2 Key Benefits

- *Real-Time Decision-Making*: The integrated system empowers farmers with real-time insights into crucial environmental parameters. By analyzing this data and applying intelligent fuzzy logic, the system can make informed decisions to meet the watering requirements of plants promptly.
- *Enhanced Security*: Blockchain technology is utilized to enhance the security of the IoT-enabled system. This ensures that only trusted devices have access to the network, mitigating potential security threats.
- *Remote Monitoring*: Multiple users can remotely monitor and interact with the system simultaneously using an Android application. This remote accessibility allows for efficient management of agricultural operations.
- *Water Efficiency*: The system not only notifies users about watering requirements but also provides alerts for turning water motors on and off. This feature contributes to efficient water resource management.

4.3.2.3 Results and Implications

The experimental outcomes of the proposed system, as demonstrated by Li Ting et al.'s research, demonstrate its scalability, efficiency, and security. It effectively monitors and optimizes climate and environmental parameters, contributing to plant health and increased agricultural productivity. While the current system focuses on eight types of plants, future directions could involve implementing advanced techniques like genetic algorithms or neural networks to provide even more precise and accurate recommendations. This case study highlights the transformative potential of Edge AI and blockchain integration in agriculture, showcasing how modern technologies can enhance efficiency, productivity, and sustainability in this vital sector.

4.3.3 Enhancing Data Collection Efficiency in Smart Agriculture through Edge Computing

The utilization of wireless sensor networks (WSNs) in agriculture has brought about noteworthy progress in data-centric farming techniques. Nevertheless, conventional agricultural WSNs frequently grapple with problems such as duplicated data and prolonged response times when confronted with critical events (CEs). These hurdles lead to amplified time and energy usage, impeding the possible advantages of precision agriculture. To tackle these predicaments, an inventive data collection technique, driven by edge computing (EC), has been introduced by Xiaomin Li et al. [13] to address CEs in the realm of smart agriculture. This case study delves into the methodology and results of this innovative approach.

4.3.3.1 Challenges Addressed

- *Data Redundancy*: Traditional WSNs generate large volumes of redundant data, leading to inefficiencies in data collection, storage, and analysis.
- *High Latency*: CEs in agriculture require real-time data collection and decision-making. High latency in data transmission can hinder timely responses, impacting crop health and yield.

4.3.3.2 Key Benefits

The proposed approach offers several key benefits:

- *Data Reduction*: By extracting and prioritizing key features data types (KFDTs) from historical datasets, the approach minimizes redundant data, optimizing data collection efficiency.
- *Low Latency*: Through EC, event types are swiftly identified based on minimum average variance, enabling rapid data collection by sensing nodes within latency constraints.
- *Resource Efficiency*: By reducing the number of required sensors, minimizing sensing time, optimizing data collection volume, and streamlining communication time, the approach enhances resource utilization efficiency.

4.3.3.3 Results and Implications

A real-world testbed within a smart greenhouse was established to validate the approach proposed by Xiaomin Li et al. The results underscored its effectiveness in reducing data redundancy, achieving low-latency data collection, and optimizing resource utilization. In comparison to conventional methods, the proposed strategy notably enhanced the equilibrium between data validity, energy consumption, and latency.

4.3.4 Enhancing Precision Agriculture with Low-Power Edge AI

In the domain of precision agriculture, characterized by the utmost importance of efficient resource utilization and crop optimization, the incorporation of AI has surfaced as a highly promising pathway to enhance farming methods. This case study explores pioneering work presented by Dmitrii Shadrin et al. [14] focusing on the detection of seed germination using AI within a low-power embedded system, offering insights into the transformative potential of Edge AI in agriculture.

4.3.4.1 Challenges Addressed

- *Low-Power AI Integration*: The integration of AI into resource-constrained embedded systems is a challenge, and this study aims to demonstrate the feasibility and benefits of such integration in agricultural contexts.
- *Efficient Seed Germination Detection*: Traditional seed germination monitoring methods may lack efficiency and accuracy. The paper addresses the need for a more effective approach using AI and computer vision.

4.3.4.2 Key Benefits

- *Resource-Efficient AI*: The study showcases the development of a low-power sensing system equipped with AI capabilities. This innovation allows for AI processing at the edge, reducing the reliance on cloud-based solutions and minimizing data transmission requirements.
- *Enhanced Germination Detection*: By implementing a Convolutional Neural Network (CNN), the system presented in this study achieves impressive results, boasting an average Intersection over Union (IoU) score of 83% on the test dataset and 97% accuracy in seed recognition on the validation dataset. Such a high level of accuracy contributes significantly to the detection of seed germination.

4.3.4.3 Results and Implications

In practice, the low-power embedded system equipped with AI capabilities can independently detect seed germination dynamics without relying on extensive data transmission to cloud servers. This not only cuts down on data-related expenses but also guarantees real-time monitoring and decision-making.

Furthermore, the scalability of this approach opens doors to widespread applications in precision agriculture. It enables the assessment of growing systems' performance, predictions about harvesting periods, and opportunities for resource optimization during the early stages of plant growth. This optimization, driven by AI insights, contributes to more efficient resource management in precision agriculture.

The study's experimental comparison between the low-power embedded system and a desktop computer highlights the advantages of Edge AI. While the desktop computer may be significantly faster, the low-power embedded system's ability to operate on battery power makes it suitable for remote and autonomous IoT applications, aligning perfectly with the emerging trends in smart agriculture.

4.3.5 Revolutionizing Precision Agriculture with Edge AI Sensor Virtualization

In the ever-changing realm of precision agriculture, effective resource management is of utmost importance. The fusion of AI and EC has opened avenues for creative solutions to tackle the issues encountered in conventional agricultural sensor-cloud systems. This case study explores the pioneering work presented by Rituparna Saha et al. [15]. The research introduces an innovative sensor-cloud architecture enhanced by EC and a sophisticated virtual sensing scheme named DLSense, aimed at transforming precision agriculture.

4.3.5.1 Challenges Addressed

- *Latency and Resource Consumption*: Traditional agricultural sensor-cloud systems suffer from high latency and substantial energy and bandwidth consumption due to the periodic transmission of raw sensor data to the cloud.
- *Unreliable Sensors*: The reliance on working sensor nodes poses limitations, especially in regions with damaged or inactive sensors.
- *Privacy Concerns*: Sharing sensitive farming data with third-party service providers raises privacy issues.

4.3.5.2 Key Benefits

- *Empowering Edge Devices*: The research introduces a modified sensor-cloud architecture that harnesses the capabilities of edge devices as an intermediary layer for sensor virtualization. This approach reduces the latency in service provisioning and decreases resource usage.
- *DLSense Virtual Sensing*: DLSense, an intelligent virtualization scheme, employs correlation theory and distributed learning within edge devices to forecast sensor data. This enables the exchange of trained models instead of raw data, ensuring data privacy while enhancing service accessibility.

4.3.5.3 Results and Implications

The performance of the DLSense scheme was assessed through thorough simulations and an experimental case study in precision agriculture. The findings underscored several notable advantages:

- *Latency and Cost Reduction*: DLSense achieves an 81% reduction in latency and a 66% reduction in service cost compared to state-of-the-art methods. This leads to quicker and more cost-effective data processing and analysis.
- *Increased Service Availability*: DLSense enhances service availability by 39%, ensuring that precision agriculture systems continue to operate effectively, even in regions with sensor failures.

This case study sheds light on the potential of Edge AI sensor virtualization in revolutionizing precision agriculture. The integration of edge devices, AI, and DLSense not only addresses the challenges of latency, resource consumption, and sensor reliability but also ensures data privacy. The promising results of reduced energy consumption, latency, and service costs, along with increased service availability, showcase the transformative impact of this approach.

4.3.6 Enhancing Crop Yield Prediction in Smart Farms with Edge AI

Crop yield prediction is a critical technique in agriculture, and the advent of smart farms, empowered by the IoT, has ushered in a wealth of real-time environmental data like temperature and humidity. However, challenges arise when handling sensitive or private data that cannot be transmitted over the internet. This case study explores innovative work presented by Junyong Park et al. [16]. The study introduces a scalable data analysis framework that leverages EC to pre-process and analyze private data, ultimately improving crop yield estimation.

4.3.6.1 Challenges Addressed

- *Private Data Handling*: Protecting sensitive farm data while still enabling effective analysis is a significant challenge, especially when internet connectivity is limited.
- *Data Separation*: Separating the data analysis process into local edge analysis and centralized server analysis while maintaining accuracy poses technical challenges.

4.3.6.2 Key Benefits

- *Scalable Data Analysis*: The proposed framework adopts a layered approach where edge devices pre-process and analyze local data, including growth data from environment sensors. This local analysis helps estimate current crop potential based on growth state.
- *Efficient Data Transmission*: By transmitting only early harvest rate information to the central server, the framework reduces the complexity of data transmission while enabling the server to trend-line predict future harvest yields.

4.3.6.3 Results and Implications

The study conducted real-world experiments in a tomato farm, demonstrating the effectiveness of the layered data analysis framework:

- *Accuracy Comparable to Server Analysis*: The results show that the error rate achieved by edge analysis is comparable to server-only analysis, even with significantly reduced feature sets.
- *Improved Complexity*: The proposed framework significantly reduces complexity while maintaining accurate results, making it a practical solution for smart farms with limited internet connectivity.

The layered data analysis method introduced in this case study offers a promising solution for enhancing crop yield prediction in smart farms. By leveraging EC for local data analysis and transmitting only essential information to the central server, sensitive data can remain protected from the internet while still contributing to accurate yield estimates.

4.3.7 Enhancing Data Sensing in Smart Agriculture through EC

A data sensing framework that leverages the synergy of EC and the IoT has been developed to address the challenges of acquiring valuable data throughout the crop lifecycle cost-effectively. This case study examines the innovative research conducted by Rihong Zhang and Xiaomin Li [17], with a specific focus on a novel data-sensing strategy aimed at enhancing the efficiency and cost-effectiveness of data collection during various crop growth stages.

4.3.7.1 Challenges Addressed

- *Limited Data Value*: Existing methods struggle to capture high-value data that correlates with specific crop growth stages.
- *Data Correlation*: Weak data correlation hinders the ability to derive meaningful insights from collected data.
- *High Data Collection Costs*: Traditional data collection approaches are resource-intensive and costly.

4.3.7.2 Key Benefits

- *Data Sensing Strategy*: The proposed strategy comprises four distinct phases:
  - Crop growth stage division using Gath-Geva (GG) fuzzy clustering.

       o Prediction of current crop growth stages employing a Tkagi-Sugneo (T-S) fuzzy neural network.

       o Optimization of environmental sensing parameters based on Deng's gray relational analysis.

       o An adaptive sensing method has been devised to accommodate sensing nodes with constraints on their effective sensing area.

- *Improved Data Value*: The strategy significantly enhances the value of collected sensing data by aligning it with specific crop growth stages.

4.3.7.3 Results and Implications

The study conducted extensive experiments to evaluate the performance of the proposed data sensing strategy:

- *High Accuracy*: The results demonstrate that the strategy can effectively divide and predict crop growth stages with high accuracy.
- *Efficiency Gains*: The strategy substantially reduces data sensing and collection times, resulting in decreased energy consumption and lower data collection costs.
- *Enhanced Data Value*: By improving the alignment of collected data with crop growth stages, the strategy elevates the overall value of sensing data.

Smart agriculture requires a robust data monitoring system capable of cost-effectively collecting valuable data throughout the entire crop life-cycle. This case study introduces a data collection framework that integrates sensors, EC, and IoT. It presents a data sensing strategy based on EC principles, consisting of four phases. This strategy aims to overcome challenges related to data value, correlation, and collection costs, ultimately yielding more efficient and valuable data.

4.3.8 Advancing Vineyard Management through Edge AI

Recent years have witnessed a notable shift in agriculture, with growing recognition of the potential benefits of Information and Communications Technology (ICT) tools and techniques. Official reports advocating for improved product quality, quantity, and economic conditions have underscored the need for the farming sector to embrace these innovations. Consequently, the deployment of sensing devices in agriculture has surged, paving the way for the emergence of smart agriculture. This evolving concept encompasses various activities, most notably field monitoring, which provides decision-making support for actions like irrigation and fertilization.

In addition to conventional sensing devices utilizing internet protocols for data transfer, the realm of smart agriculture incorporates crop models. These models harness sensor data to deliver enhanced insights and recommendations for farmers, ultimately enhancing the quality and quantity of agricultural production.

This case study delves into the innovative work presented by Sergio Trilles and his team [18]. The study centers on the development of a cost-effective sensorized platform aligned with the IoT paradigm. The platform focuses on monitoring meteorological conditions, primarily to aid in the detection of mildew disease in vineyards. Leveraging EC and drawing inspiration from advances in Geographic Information Science (GIScience), this platform was deployed in a vineyard located in Vilafamés, Spain.

4.3.8.1 Challenges Addressed

- *Cost Reduction*: The platform aims to optimize the timing of downy mildew disease control treatments, thereby minimizing the costs associated with unnecessary chemical applications.
- *Environmental Impact*: By accurately timing treatment applications, the platform contributes to reducing the ecological footprint by limiting chemical usage and soil contamination.
- *Labor Efficiency*: The system enhances labor efficiency by eliminating the need for unnecessary vineyard visits and treatments.

4.3.8.2 Key Benefits

- *Cost Savings*: The platform identifies the optimal moments to apply downy mildew control treatments, resulting in significant reductions in chemical expenses.
- *Environmental Sustainability*: By curbing chemical usage, the platform promotes a healthier environment and more sustainable agricultural practices.
- *Labor Optimization*: Farm labor is used more efficiently, as treatments are only administered when needed, freeing up working hours for other essential tasks.

4.3.8.3 Results and Implications

The sensorized platform, referred to as SEnviro, has demonstrated its effectiveness in monitoring vineyards. By harnessing IoT and EC paradigms, an enhanced version of this platform processes data locally, enabling immediate information computation upon data collection. This approach not only reduces data transmission time and energy consumption but also enhances decision-making accuracy.

The primary focus of the platform's application is the timely detection of downy mildew treatment opportunities. Through real-time analysis of meteorological data, the system offers valuable insights for disease control. While this case study primarily tested the IoT platform's functionality and did not aim to validate disease models, the results were promising.

4.3.9 Transforming Animal Welfare Monitoring in Agriculture with Edge AI Computing

Recent years have witnessed a significant shift in agriculture, driven by the emergence of smart sensing and computing technologies. Smart agriculture systems, while still relatively nascent, hold immense promise for revolutionizing crop and livestock farming practices. This case study, inspired by the research conducted by Marcel Caria and his team [19], focuses on the development of an open and cost-effective smart farm computing system for monitoring animal welfare. The system leverages EC, fog computing, and cloud connectivity to create a comprehensive solution that addresses the complex and multifaceted aspects of animal well-being.

4.3.9.1 Challenges Addressed

- *Limited Accessibility*: Many existing smart agriculture systems are closed for experimentation, limiting opportunities for innovation and customization, particularly in the domain of animal welfare monitoring.
- *Comprehensive Monitoring*: Animal welfare encompasses a wide range of parameters and factors, making it challenging to develop a holistic monitoring solution that caters to the diverse needs of stakeholders.
- *Cost-Efficiency*: Keeping costs low while ensuring effective monitoring of animal welfare is a critical consideration for the agricultural sector.

4.3.9.2 Key Benefits

The proposed smart farm computing system offers several key advantages:

- *Open and Low-Cost*: By adopting open-source principles and utilizing cost-effective Raspberry Pi devices as edge nodes, the system provides accessible and customizable solutions for animal welfare monitoring.
- *Fog Computing Layer*: The conceptual creation of a fog computing layer enables real-time data processing at the edge, reducing latency and enhancing responsiveness in monitoring various parameters.
- *Cloud Connectivity*: Seamless integration with cloud computing systems allows for centralized data storage, analysis, and remote access, providing a comprehensive overview of animal welfare.
- *Multi-parameter Monitoring*: The system effectively monitors and collects data on multiple parameters relevant to animal welfare, offering valuable insights to stakeholders.

4.3.9.3 Results and Implications

The proposed smart farm computing system, which combines EC with cloud connectivity, demonstrates its effectiveness in monitoring diverse parameters related to animal welfare. By leveraging Raspberry Pi devices as edge devices, the system creates a fog computing layer that processes data locally, ensuring real-time responsiveness.

This innovative approach to animal welfare monitoring not only addresses the immediate needs of farmers and livestock caretakers but also offers possibilities for broader stakeholder engagement. Animal welfare, a multifaceted concept, can be measured, collected, evaluated, and shared, creating opportunities for improving the well-being of animals and fostering high-tech innovations in the agricultural sector.

This case study showcases the transformative potential of Edge AI computing in the context of smart agriculture, specifically in monitoring and enhancing animal welfare. By providing accessible, cost-effective, and open solutions, it opens up new avenues for improving agricultural practices, promoting animal well-being, and driving technological advancements in the sector.

4.3.10 Farming with CareBro: An Edge AI–Enabled IoT Farming Assistant

In the wake of the post-COVID-19 era, agriculture faces a pivotal transformation. The new imperative lies in enhancing productivity and ensuring the safety of agricultural produce, all the while reducing reliance on traditional labor-intensive practices. Technology emerges as the beacon of hope, and a contactless, reliable, and secure approach is the need of the hour. "CareBro" is a ground-breaking solution designed to redefine and revolutionize farming practices. This IoT-based smart agriculture system, powered by EC, takes the reins of farm management, making it autonomous and remote. It integrates seamlessly with smart farm sensors, orchestrating activities such as crop yield optimization, ethical pest management, and precision irrigation. CareBro operates as a cutting-edge solution, connected to the farmer via the cloud, offering real-time monitoring and intelligent decision-making. This innovation, authored by Atharv Tendolkar and Ramya [20], represents an innovative farm management solution, applicable across diverse farming landscapes, from urban and rural settings to large-scale and small-scale operations.

4.3.10.1 Challenges Addressed

- *Labor Shortage*: CareBro steps in as a solution for farming challenges exacerbated by labor shortages, ensuring the continuity of farm operations even during crises.
- *Productivity Enhancement*: The system aims to significantly increase crop yields while ensuring sustainable and ethical farming practices.
- *Precision Agriculture*: CareBro tackles the need for precision in pest management and irrigation, optimizing resource usage and minimizing environmental impact.
- *Remote Monitoring*: With real-time monitoring and remote control, it bridges the gap between farmers and their fields, offering a layer of convenience and control.

4.3.10.2 Key Benefits

- *Maximum Yield*: CareBro empowers farmers to maximize crop yields while providing valuable insights into challenges and mitigation strategies, all managed remotely and securely.

- *Adaptability*: The system is designed for adaptability, with sensors calibrated through beta testing to respond effectively to changing environmental conditions.
- *Multi-Functionality*: CareBro's potential extends beyond farm management; it can serve as a central hub for various farming services, including automated tasks like combine harvesting, nutrient spraying, and irrigation.
- *EC*: With its IoT network grid and local EC capabilities, CareBro enhances intelligence at every node across the entire farm.

4.3.10.3 Results and Implications

The development of CareBro has showcased its potential to revolutionize farming practices. Beta testing and sensor calibration have ensured the device's adaptability to dynamic environments. Integrating multiple modules and enhancing connectivity through a physical chassis resembling a scarecrow has been a crucial step. CareBro's scalability and multi-functionality, capable of serving as a central hub for various farm services, hold significant promise for the future. With its wide-ranging applicability, economic viability, and potential for scalability, CareBro emerges as a quintessential farm assistant for the modern, IoT-driven farming landscape.

CareBro operates sustainably, offering convenience and precision in farm management, fostering trust in technology. As technology continues to evolve, with the advent of 5G and advancements in EC, CareBro is poised to meet farmers' evolving needs and expectations. It stands as a testament to customer-centricity, sustainable farming practices, ethical principles, and safety.

4.4 DISCUSSION

The case studies presented in this chapter offer a rich tapestry of insights into the world of Edge AI applications in agriculture. These innovative systems bring to the forefront a multitude of functional features, utilization scenarios, benefits, socio-economic impacts, and adoption considerations that collectively define their profound significance to stakeholders in the agricultural domain.

4.4.1 Functional Features and Utilization

- *Real-Time Monitoring*: Edge AI systems excel in real-time monitoring of crucial agricultural parameters. This feature is indispensable in modern farming, as it allows farmers to obtain up-to-the-minute information on soil conditions, weather patterns, and crop health. The utilization of sensors, cameras, and data analytics enables the continuous collection and interpretation of data. As a result, farmers can make prompt decisions regarding irrigation, pest control, and disease management.
- *Data-Driven Decision-Making*: These systems leverage advanced analytics and machine-learning algorithms to generate data-driven recommendations. By analyzing large datasets collected from farms, they provide actionable insights for precise resource allocation. For example, they can determine the exact amount of water required for irrigation, the optimal timing for pesticide application, or the need for specific nutrients. This data-driven decision-making not only enhances crop yields but also minimizes the environmental impact of farming practices.
- *Interconnectivity*: Accessibility is a cornerstone of Edge AI systems. They are designed for seamless interconnectivity, allowing farmers to monitor and manage their farms remotely. Mobile applications and web interfaces provide user-friendly access to real-time data and control capabilities. Farmers can receive alerts and notifications, view reports, and adjust settings through these interfaces, regardless of their physical location. This functionality enhances convenience and efficiency in farm management.
- *Security and Privacy*: Data security and privacy are paramount. To address these concerns, some systems incorporate blockchain technology. Blockchain ensures the integrity and security of data by creating tamper-proof records. Furthermore, these systems often prioritize local data processing at the edge, meaning that sensitive information is processed on-site rather than being transmitted to centralized servers. This approach reduces the risk of data breaches during transmission.

4.4.2 Benefits

- *Resource Efficiency*: Edge AI systems are champions of resource efficiency. By continuously monitoring soil moisture, temperature, humidity, and other parameters, they optimize resource utilization.

This optimization extends to water, fertilizers, pesticides, and energy. For instance, the precise control of irrigation systems reduces water wastage, which is particularly critical in regions facing water scarcity. Similarly, tailored fertilization and pest management reduce chemical runoff, contributing to environmental sustainability. These efficiency gains result in cost savings for farmers.

- *Increased Crop Yield*: Timely interventions facilitated by Edge AI systems translate into increased crop yields. By detecting diseases, pests, or water stress at an early stage, these systems enable farmers to take swift corrective actions. For instance, they can trigger irrigation when soil moisture drops below a certain threshold, preventing crop wilting. Moreover, they can identify the presence of pests or diseases through image recognition, allowing for targeted treatments. The cumulative effect of these measures is higher crop productivity, ultimately benefiting both food security and farm profitability.

- *Data-driven Insights*: Edge AI systems provide farmers with valuable data-driven insights. As climate patterns become more unpredictable and farming practices evolve, these insights are crucial for making informed decisions. Farmers gain access to historical and real-time data, allowing them to adapt to changing conditions. For example, they can adjust planting schedules based on weather forecasts or optimize irrigation regimes according to soil moisture levels. This data-driven approach fosters resilience in agriculture, enabling farmers to navigate challenges effectively.

4.4.3 Socio-economic Impact

- *Empowering Smallholders*: In regions characterized by a high prevalence of smallholder farmers, Edge AI systems play a pivotal role in leveling the playing field. These technologies provide access to advanced farming practices and tools that were once beyond the reach of many small-scale farmers. As a result, rural communities benefit from enhanced agricultural productivity and improved livelihoods. The socio-economic impact is particularly significant in developing countries where agriculture is a primary source of income.

- *Environmental Sustainability*: The resource-efficient practices promoted by Edge AI systems contribute to environmental sustainability. Reduced water usage, minimized chemical runoff, and lower energy consumption collectively mitigate the environmental impact of agriculture. These systems align with global sustainability goals by promoting eco-friendly farming practices. The socio-economic implications extend to cleaner environments, healthier ecosystems, and a reduced carbon footprint.

4.4.4 Adoption and Future Outlook

- *Scalability*: One of the key drivers of adoption is the scalability of Edge AI systems. They are designed to accommodate multiple users and farms simultaneously. This scalability makes them accessible to a broad spectrum of stakeholders, from individual farmers to large agricultural enterprises. As more farms adopt these technologies, the collective impact on agriculture becomes increasingly significant.
- *Interdisciplinary Integration*: Edge AI systems demonstrate versatility through interdisciplinary integration. They incorporate various technologies, including IoT, blockchain, and cameras for pest and plant health monitoring. This adaptability positions them for further growth and adaptation to evolving agricultural challenges. For instance, the integration of cameras for pest detection and plant health monitoring expands their utility beyond resource management to include proactive pest control.

4.5 CONCLUSION

In this chapter, a journey through the dynamic landscape of Edge AI in agriculture was embarked upon. The fundamental concept of Edge AI was explored, its potential was deciphered, and its pivotal role in reshaping modern agriculture was recognized. A new era of immediacy, responsiveness, and efficiency in farming practices is heralded by the integration of AI with EC and the IoT.

As a series of case studies was delved into, how Edge AI has been applied to address the most pressing challenges in agriculture was witnessed, spanning both national and international contexts. The transformative potential of Edge AI was demonstrated by these real-world examples, showcasing its versatility and adaptability. From precision agriculture and resource optimization to livestock management and crop monitoring, Edge AI was proven to be a catalyst for innovation, enabling farmers and stakeholders to make data-driven decisions, reduce waste, and enhance productivity.

The adoption of Edge AI in agriculture is not without its challenges, including issues related to connectivity, data privacy, and scalability. However, solutions are met that ensure seamless integration and the safe-guarding of sensitive agricultural data.

In summary, a powerful ally in the pursuit of sustainable and efficient farming practices is Edge AI. Stakeholders are empowered with real-time insights, enabling informed decisions that benefit both the agricultural industry and the environment. As we move forward, the potential for innovation in Edge AI remains boundless, promising a future where agriculture becomes smarter, greener, and more resilient.

REFERENCES

[1] W. Shi, J. Cao, Q. Zhang, Y. Li, and L. Xu, "Edge computing: Vision and challenges," *IEEE Internet Things J.*, vol. 3, no. 5, pp. 637–646, 2016.

[2] Y. Mao, C. You, J. Zhang, K. Huang, and K. B. Letaief, "A survey on mobile edge computing: The communication perspective," *IEEE Commun. Surv. Tutor.*, vol. 19, no. 4, pp. 2322–2358, 2017.

[3] D. Situnayake and J. Plunkett, *AI at the edge: Solving real-world problems with embedded machine learning*. Sebastopol, CA: O'Reilly Media, 2023.

[4] Z. Zhou, X. Chen, E. Li, L. Zeng, K. Luo, and J. Zhang, "Edge intelligence: Paving the last mile of artificial intelligence with edge computing," *Proc. IEEE Inst. Electr. Electron. Eng.*, vol. 107, no. 8, pp. 1738–1762, 2019.

[5] T. H. Luan, L. Gao, Z. Li, Y. Xiang, G. Wei, and L. Sun, "Fog computing: Focusing on mobile users at the edge," *arXiv* [cs.NI], 2015.

[6] J. Wang, Z. Feng, S. George, R. Iyengar, P. Pillai, and M. Satyanarayanan, "Towards scalable edge-native applications," in *Proceedings of the 4th ACM/IEEE Symposium on Edge Computing*, 2019.

[7] F. J. Ferrández-Pastor, J. M. García-Chamizo, M. Nieto-Hidalgo, J. Mora-Pascual, and J. Mora-Martínez, "Developing Ubiquitous Sensor Network platform using Internet of Things: Application in precision agriculture," *Sensors (Basel)*, vol. 16, no. 7, p. 1141, 2016.

[8] S. T. Oliver, A. González-Pérez, and J. H. Guijarro, "An IoT proposal for monitoring vineyards called SEnviro for agriculture," in *Proceedings of the 8th International Conference on the Internet of Things*, 2018.

[9] D. Berckmans, "General introduction to precision livestock farming," *Anim. Front.*, vol. 7, no. 1, pp. 6–11, 2017.

[10] X. Zhang, Z. Cao, and W. Dong, "Overview of edge computing in the agricultural internet of things: Key technologies, applications, challenges," *IEEE Access*, vol. 8, pp. 141748–141761, 2020.

[11] H. T. Minh, T. P. Anh, and V. N. Nhan, "A novel light-weight DCNN model for classifying plant diseases on internet of things edge devices," *Mendel*, vol. 28, no. 2, pp. 41–48, 2022.

[12] L. Ting, M. Khan, A. Sharma, and M. D. Ansari, "A secure framework for IoT-based smart climate agriculture system: Toward blockchain and edge computing," *J. Intell. Syst.*, vol. 31, no. 1, pp. 221–236, 2022.

[13] X. Li, Z. Ma, J. Zheng, Y. Liu, L. Zhu, and N. Zhou, "An effective edge-assisted data collection approach for critical events in the SDWSN-based agricultural internet of things," *Electronics (Basel)*, vol. 9, no. 6, p. 907, 2020.

[14] D. Shadrin, A. Menshchikov, D. Ermilov, and A. Somov, "Designing future precision agriculture: Detection of seeds germination using artificial intelligence on a low-power embedded system," *IEEE Sens. J.*, vol. 19, no. 23, pp. 11573–11582, 2019.

[15] R. Saha, A. Chakraborty, S. Misra, S. K. Das, and C. Chatterjee, "DLSense: Distributed learning-based smart virtual sensing for precision agriculture," *IEEE Sens. J.*, vol. 21, no. 16, pp. 17556–17563, 2021.

[16] J. Park, J.-H. Choi, Y.-J. Lee, and O. Min, "A layered features analysis in smart farm environments," in *Proceedings of the International Conference on Big Data and Internet of Thing*, 2017.

[17] R. Zhang and X. Li, "Edge computing driven data sensing strategy in the entire crop lifecycle for smart agriculture," *Sensors (Basel)*, vol. 21, no. 22, p. 7502, 2021.

[18] S. Trilles, J. Torres-Sospedra, Ó. Belmonte, F. J. Zarazaga-Soria, A. González-Pérez, and J. Huerta, "Development of an open sensorized platform in a smart agriculture context: A vineyard support system for monitoring mildew disease," *Sustain. Comput. Inform. Syst.*, vol. 28, no. 100309, p. 100309, 2020.

[19] M. Caria, J. Schudrowitz, A. Jukan, and N. Kemper, "Smart farm computing systems for animal welfare monitoring," in *2017 40th International Convention on Information and Communication Technology, Electronics and Microelectronics (MIPRO)*, 2017.

[20] A. Tendolkar and S. Ramya, "CareBro (personal farm assistant): An IoT based smart agriculture with edge computing," in *2020 Third International Conference on Multimedia Processing, Communication & Information Technology (MPCIT)*, 2020.

5

Crop Recommender: Machine Learning–Based Computational Method to Recommend the Best Crop Using Soil and Environmental Features

Syed Nisar Hussain Bukhari, Jewiara Khursheed Wani,

Ummer Iqbal, and Muneer Ahmad Dar

National Institute of Electronics and Information Technology (NIELIT), Srinagar, India

5.1 INTRODUCTION

In our country, agriculture is one of the main sectors on which a large number of people are dependent [1]. In spite of having a great variety of topographic characteristics, landforms, and climatic zones that support various types of cultivation, the yield obtained is still less due to field mismanagement and the erroneous selection of crops [2]. To adequately nourish the rapidly expanding population, it is crucial to prioritize the proportional increase in crop yield. This can be accomplished by addressing the underlying causes of reduced crop production, including insufficient planning, mismanagement of livestock, and the unpredictable nature of weather conditions. Since the complicated phenomenon of crop production is impacted by agro-climatic input parameters, therefore the specifications for agricultural inputs vary according to the farmer and field [3]. In addition to the various factors that affect crop production, another mammoth problem that agriculturists are facing is the rise of global warming. Global warming being on peak leads to droughts and non-predictable climatic conditions. Moreover, farmers working in the traditional ways lead to the problem of field mismanagement and hence affect productivity. According to the reports, due to unpredictable climatic conditions over the years, the farmer's income has reduced by 20.25% [4].

Agriculture holds a significant position in India's gross domestic product (GDP). The crucial factor for achieving a high crop yield is the proper selection of appropriate crops at the right time of year. However, farmers often rely on market trends and ancestral instincts, which have proven to be ineffective methods [5]. Adapting the proposed methodology will help farmers overcome the problem of scarcely quantifiable yield and contribute more and more toward the country's GDP [6]. The use of algorithms will also aid in reaching an efficient, cost-effective solution to the difficulties faced by the farmers. The algorithms trained and tested will be used for crop recommendation based on various inputs like Nitrogen, Phosphorus, Potassium (NPK) content and pH of soil, temperature, and humidity of the area. The predicted crops included in this methodology are *maize, cotton, rice, coffee, pigeon peas, moth beans, jute, chickpea, coconut, papaya, orange, apple, muskmelon, watermelon, grapes, kidney beans, banana, pomegranate, lentil, black gram, mung bean,* and *mango.*

Some of the primary novel contributions of this chapter are the following:

- In this study, a system has been proposed that will enable agriculturists to select accurate crops based on environmental factors and soil parameters.
- The proposed system can reduce the erroneous methods applied during the selection of a crop, which in turn can prove helpful in improving crop productivity.
- A crop recommender (CR) tailored to provide recommendations for the most suitable crop to plant will be developed.

Four sections make up this chapter. In the first section, an introduction has been presented, and various issues related to decreased crop yield have been discussed. The second section contains an account of related work done in this field. An experimental scenario has been covered in the third section, which includes the dataset used and methods employed. Results have been provided in the fourth section. Lastly, a conclusion of the study has been drawn.

5.2 RELATED WORK

The work completed in the area of crop recommendation by various research-ers is included in this section. A thorough description of the study project is provided in the remaining portion of this section, and the summarization of this section is given in Table 5.1. Doshi et al. [1] developed Agro Consultant, which takes into consideration all the relevant attributes – i.e., temperature, location, rainfall, and soil condition – for predicting crop appropriate-ness using four algorithms – namely, DT, Random Forest (RF), NN, and K-nearest neighbor. This study also presented a rainfall predictor (a secondary system) that enables early predictions of rainfall [1]. Kumar et al. [7] built a model called a selection of crops, which is based on parameters like weather, soil type, water density, and crop kind. This method employed models like RF, support vector machine (SVM), DT, KNN, and Regularized Greedy Forest (RGF) [7]. Kumar Rajak et al. [8] employed a method that was trained on a soil database obtained from a

TABLE 5.1

Related Work

Author	Year	Description	Algorithm
Doshi et al.	2018	Prediction of crop suitability using all relevant attributes (i.e., temperature, rainfall, and soil condition)	Decision tree, NN, RF, and KNN
Kumar et al.	2015	Selection of crops using parameters soil type, water density, and crop kind	RF, SVM, decision tree, KNN, and RGF
Kumar Rajak et al.	2017	Crop recommendation using a validated dataset for maximizing crop yield	Bagging ANN, SVM, Naive Bayes, RF
Savla et al.	2015	Comparative analysis using various classification algorithms, particularly on soybean crops	SVM, NN, RF, REP Tree
Shakil Ahamed et al.	2015	Data mining techniques were conducted for the cereal crops	KNN, linear regression, ANN
Suresh et al.	2021	Developed a system called efficient crop yield, which predicts the suitability of crops by employing two datasets (i.e., datasets of crop and particular location)	SVM

farm. The data retrieved after testing from the lab was then provided to a recommendation system based on an ensemble model of Artificial Neural Network (ANN) and SVM [8]. Savla et al. [9] made a relative analysis of classification algorithms and analyzed the rate of prediction in precision agriculture. The algorithms employed in the study were RF, NN, SVM, Reduced Error Pruning Tree (REP Tree), Bagging, and Bayes. Among the aforementioned algorithms, the bagging algorithm drew the best result for predicting the yield [9]. Shakil et al. [10] specified the use of data mining techniques that can help government organizations and agriculturists draw out the necessary information to reach better conclusions. This ultimately results in increased production levels. These techniques were particularly applied for estimating cereal crops in Bangladesh. The algorithms employed in their study included clustering and classification using KNN, Linear Regression, and ANN [10]. Suresh et al. [11] developed an "Efficient Crop Yield Recommendation System" based on SVM that recognizes a specific crop in accordance with the given data. The system recommended a specific crop in accordance with the parameters (NPK and pH) values [11].

5.3 DATA AND METHODS

5.3.1 Experimental Scenario

For experimentation, the programming environment python is adopted. Bagging models have been used for experimentation. The steps of the experiment are listed in Figure 5.1.

The steps undertaken in this study are as follows: procure the relevant dataset, preprocess the obtained dataset, visualize the dataset, build the ML model, and evaluate the same. The data is preprocessed and then fed to models for classification. Lastly, the comparative analysis between the ensemble of homogenous classifiers and single classifiers is conducted.

5.3.2 Dataset Collection

The data used for the study is a combination of two datasets procured from Kaggle [12] crop recommendation and *agric dataset*, which consist of 2,202 and 4,100 samples, respectively. After merging the samples, the resultant dataset was divided into 7:3 portions for train and testing.

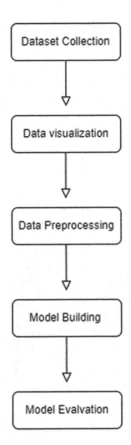

FIGURE 5.1
Experimental steps.

5.3.3 Data Visualization

Data visualization plays a crucial role in simplifying intricate information and making it comprehensible. To assess the correlation among various features, a heat map has been utilized. The heat map indicates a significant positive correlation between phosphorous (P) and potassium (K) while revealing a negative correlation between phosphorous (P) and rainfall, pH and potassium (K), and nitrogen (N) and temperature, as presented in Figure 5.2. "Univariate Analysis." After comprehending the correlations among the features, this data becomes instrumental in making informed decisions for selecting the most suitable crop. Following the univariate analysis by the bar graph aids in identifying the specific parameters needed for the particular crop, as depicted in Figure 5.3.

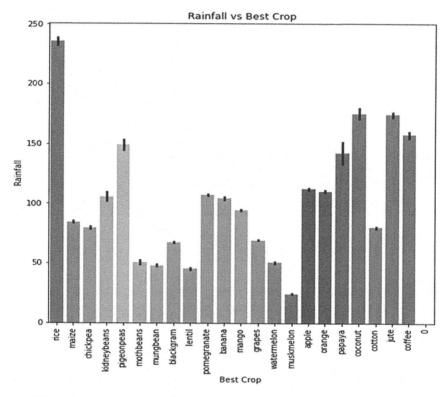

FIGURE 5.2
Univariate analysis.

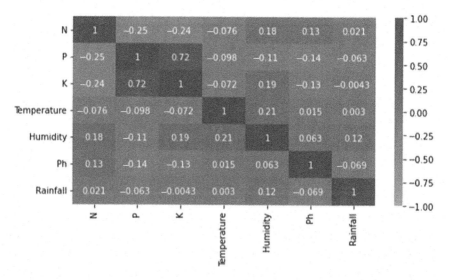

FIGURE 5.3
Heat map.

5.3.4 Data Cleaning

In the discipline of ML, data cleaning (DC) is an essential and necessary step that precedes the modeling process. This method involves identifying and rectifying errors, discrepancies, and inaccuracies present in a dataset. This process encompasses various tasks, including addressing missing values, eliminating duplicates, standardizing data formats, and handling outlier values. Adhering to this step is essential; otherwise, it can result in the creation of an inaccurate model [13].

Allowing the model to learn from pertinent or relevant data ultimately leads to improved classification outcomes [14]. After merging the datasets [12], a few null entries present in the data were removed.

5.3.4.1 Outlier Handling

One of the main parts of data preprocessing is the outlier and noise handling: fault in data collection, error during data-entering, unknown encoding, inconsistent data, inconsistent formats, etc., correspond to different types of noise and outliers [15]. It's very significant to handle it before modeling, as it affects the statistical analysis and training process of ML algorithms, leading to lower accuracy. There are various visualization and mathematical techniques by which we can detect and remove outliers, such as boxplots, z-scores, and interquartile range (IQR). In the proposed model, the outliers were detected by IQR and were handled by a capping technique.

5.3.4.2 Feature Scaling

During the preprocessing, feature scaling (FS) also counts among the important steps. It is important to bring all the features into the same scale in order to enable the model to learn from all the features equally without leading to bias. FS also helps algorithms to achieve the minimal cost function. Scaling plays a very significant role in making the ML model either a weak model or a better one [16]. FS is achieved by adopting methods like normalization and standardization. In this study, the FS was done by normalization method, as depicted in Equation (5.1) [16].

$$X_{\text{norm}} = \frac{X - \min(X)}{\max(X) - \min(X)} \qquad (5.1)$$

5.3.5 Model Building

The basic steps of building an ML model are as follows:

    i. Selection of the ML model
   ii. Fitting of the ML model
  iii. Validating ML model

Selecting an ML model is the process in which a single ML model is selected out of various possible candidates for training a dataset [17]. Fitting an ML model involves training it on a dataset to find the best parameters that capture patterns in the data. The model's parameters are adjusted to minimize the difference between predicted and actual values, aiming to create an accurate predictor for new data. The functionality of an ML model to predict values and perform classifications is achieved by generalizing the model to new datasets [18].

Validating an ML model is a crucial and important process that is performed after training a model, where the trained model is assessed by testing data. Validating the ML model's outcome is necessary to ensure its accuracy [18].

5.3.5.1 Methods

The technique proposed in this study is the ensemble technique (ET). The main aim of using ET in this study is to overcome the issues of simple modeling of one classifier and to achieve better accuracy [17]. The ET involves the integration of multiple base learners, where every base learner contributes to the outcome leading to the highest level of accuracy and reduced error of prediction. ET is particularly used to help with decision-making problems.

One type of ET is bagging, which follows the process of sub-sampling training data in order to improve the outcome of a single type of classifier. The method that is used to subsample the data is bootstrapping. The bootstrapping process involves taking samples with the replacement from a dataset and calculating a statistic (mean, median) for each sample. So this process helps to get a distribution of results for statistics of interest by which we can estimate the standard errors and confidence intervals, which eventually leads to getting a less biased estimate. The output gathers the prediction of every learner, and with the help of majority voting, the desired output is achieved [19]. Figure 5.4 depicts the ET technique used.

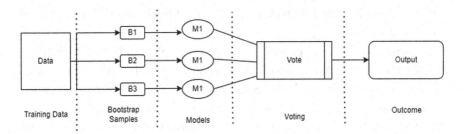

FIGURE 5.4
Bagging technique.

5.3.5.2 Technique

The first model used in this study is DT. The main reason for selecting this algorithm is that it can handle different types of data and does not require much data preprocessing [20]. Mostly, DT is used for classification tasks since DT simulates human-level thinking, which offers a straightforward way to analyze data and derive meaningful interpretations. This model involves building a treelike structure for the entire dataset and processing a single outcome at each leaf node [21]. The problems associated with training only one DT can be tackled by using the bagging model of DT, which leads to attaining better accuracy.

Deep learning is widely used in recommendation systems in order to enhance the sparse data and scalability of the method. This measures the enhancement in the quality of recommendation outcomes.

The second model trained in this study is the Neural Network (NN). The main criterion for selecting this algorithm is that it does a wondrous job of making the machines able to make decisions similar to the human brain and hence leads to better predictions by understanding the complex relationship between input and output data [22]. In this study, a homogenous bagging model was built.

5.3.5.2.1 Decision Tree

The DT algorithm is a straightforward as well as adequate supervised learning method in which the sample points are continuously divided depending on several parameters. The components of DT comprise the root node, some branches, and leaf nodes. The interior nodes in the structure trees provide descriptions related to different test scenarios.

This algorithm can be envisaged as a graphical treelike structure that predicts the output by employing a variety of specialized factors [21].

Figure 5.5 depicts the first technique. The working of the DT is illustrated in Algorithm 5.1.

Assume S to be a collection of training samples, each of which has a known class label. One feature is used to decide the class of training samples. Let's assume there are n classes. For $I = 1,... n$, let S consist of s_i samples of class C_i for $I = 1,...n$. A random sample has a probability of s_i/s that it belongs to the class C_i, where s is the number of total samples in the collection S. The anticipated information required to classify a given sample [3] is represented in Equation (5.2) [3].

$$I(s,s,...,s_n) = -\sum_{i=1}^{n} \frac{S_i}{S} \log \frac{S_i}{S} \tag{5.2}$$

Feature B with values $\{b_1, b_2, ..., b_v\}$ can be employed to split S into the subsets $\{S_1, S_2, ..., S_v\}$, where S_j consists of those samples in S that have a value bj of B. Let Sj consist of sij examples of class Ci. The entropy of B is the predicted information based on this division by B. It is the weighted average that is represented in Equation (5.3) [3]:

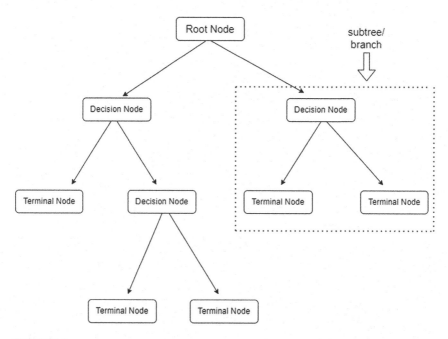

FIGURE 5.5
Decision tree.

ALGORITHM 5.1   DECISION TREE

Input – Supervised Data of N instances
  For each sample in the dataset:

  1: Examine each base scenario.
  2: For every feature *b*: evaluate the standardized information gained from splitting on *b*.
  3: Let *b_best* be the feature with higher standardized information gain.
  4: Build a decision *node* that splits on *b_best*.
  5: Recurse on the subtree attained by splitting on *b_best*, and add those nodes as leaf nodes.

Return Classifier

$$E(B) = \sum_{j=1}^{v} \frac{S^1 j + \ldots + S_n j}{S} I\left(Sj + \ldots + S_n j\right) \tag{5.3}$$

This portioning on B results in an information gain that is represented in Equation (5.4) [3].

$$\text{Gain}(B) = I\left(s, s, \ldots, S_n\right) - E(B) \tag{5.4}$$

For the relevance analysis, the information gain is computed for every attribute present in sample S. The attribute that exhibits the greatest information gain is regarded as the most distinguishing characteristic within the set. By calculating the information gain for each attribute, a ranking of the attributes is achieved, which can be used for relevance analysis to select the attributes to be used in concept description.

5.3.5.2.2 Neural Network

The nonlinear mathematical model created by mocking the biological brain is known as NN or ANN. ANNs are capable of processing nonlinear datasets and mapping them to the results. The most popular ANN for solving nonlinear datasets is multilayer perceptron (MLP). ANN consists of three main layers: input layer, hidden layer, and output layer. They are built on a network of linked nodes known as neurons. Following that, connections between these neurons are used to send signals. As learning progresses, the weights assigned to the neurons and connections are updated and modified, which boosts their performance [23], as shown in Figure 5.6. NN presents the second technique. The working of the DT is illustrated in Algorithm 5.2.

Each neuron in the receiving layer (input layer) passes information and is transmitted to connected neurons in the hidden layer. Each hidden layer unit computes the output by adding the afflicted input signals that are employed in the activation function (AF) by applying Equation (5.5) [24].

$$h(t) = f\left(\sum x(t) W_H + B_H\right) \qquad (5.5)$$

The output signal is calculated by adding the input signal and its AF using Equation (5.6) [24].

$$O(t) = f\left(\sum h(t) Wo + Bo\right) \qquad (5.6)$$

AFs are employed to each neuron present in the network. It is a mathematical "gate" between the receiver layer serving the current neuron and its result passing to the next layer. AFs can be linear and nonlinear.

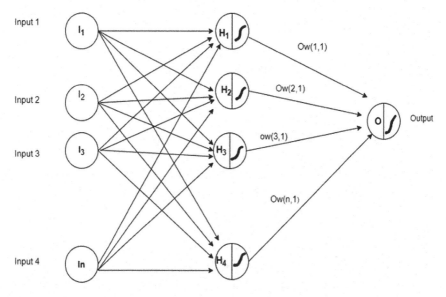

FIGURE 5.6
Neural network.

ALGORITHM 5.2   NN

Input – Supervised Data of N instances
For each sample in the dataset:

1: Initialization: Input each row of observations on the input layer and every feature at each input node. Initialize the weight values randomly but not 0.
2: Forward-Propagation: From left to right, compute and activate each node of layers with the weights controlling the impact of corresponding neurons.
3: Cost Computation: Measure the error with the predicted value and the actual value.
4: Back Propagation and Updates: From right to left, update the weights with the gradient descent to lower the error until optimization.
5: Prediction: Predict the value for test data.

Return Classifier

Nonlinear AFs are employed in the case of complex data to predict the outcome. Among the AFs, ReLU (Rectified Linear Unit) is a diversely employed activation function. The inputs with a value less than zero are mapped to zero and then mapped to one for the inputs with a value greater than zero. The best option for multiclass classification is the Softmax function, as it transforms the results into a normalized probability distribution. The outcomes of the NN pass along to the Softmax (AF) that transforms the results into probabilistic values, which finally are compiled up to one.

Softmax output function can be expressed in Equation (5.7) [24]:

$$\sigma = \frac{e^{zi}}{\sum_{j=1}^{k} e^{zj}} \tag{5.7}$$

In feed-forward NN, back propagation is the most frequently employed algorithm for training the multilayer NN, where the signals from the input layer are transmitted onward, and errors are transmitted backward to alter the weights in such a way that minimizes the output error. This process is replicated during training, and each iteration is known as an epoch. The error calculation in a back-propagation algorithm is presented in Equation (5.8) [24].

$$E = \frac{1}{2}\sum(t_k - O_k)^2 \tag{5.8}$$

Each weight is reconditioned as expressed in Equation (5.9) [25]:

$$wi = wi + \Delta wi, \tag{5.9}$$

where Δwi is the rectified factor calculated as represented in Equation (5.10) [25].

$$\Delta wi = \eta \delta jx_{ij} \tag{5.10}$$

5.3.6 Model Evaluation

The process of assessing the performance of a model is called model evaluation. Since predicting the best crop is a job that is affiliated with multiclass classification, there are four possible outputs – i.e., True Negative (TN), True Positive (TP), False Negative (FN), and False Positive (FP) [25].

For assessing the performance of a model, various metrics have been employed – i.e., accuracy, specificity, sensitivity, and area under receiver operator characteristic (AUROC) curve [26]. These metrics are defined in terms of the aforementioned results. To assess the stability and versatility of the proposed system, a method has been employed for validating the model, which is known as K-fold cross-validation (KFCV). The measures employed are defined as follows.

- Accuracy (Acc) = $\dfrac{TP + TN}{TP + TN + FP + FN}$
- Sensitivity (Sens) = $\dfrac{TP}{TP + FP}$
- Specificity (Spec) = $\dfrac{TP}{TP + FN}$
- F1 score = $\dfrac{2*Precision*Recall}{Precision + Recall}$

5.3.6.1 AUROC Curve

An essential assessment statistic for classification problems is the AUROC curve. The AUROC curve is a crucial evaluation metric used to assess the performance of a multiclass classification problem. It is widely utilized to check and visualize how well a classification model performs. The curve successfully distinguishes between noise and signal by plotting True Positive Rate (TPR) vs. False Positive Rate (FPR) at various levels. Among the values, the value present at the top-left side of the curve is considered the optimum value. Figure 5.7 shows the AUROC curve [26].

5.3.6.2 K-Fold Cross-Validation

The process of assessing the ML models on small samples of data is known as cross-validation. The technique utilizes a parameter called *k*, which determines the number of groups into which the data has been divided. For every iteration, *k*–1 subsamples are employed for training the model, and the residual part is employed for assessing the performance of the model. Finally, the results from *k*-iterations are added, and the model's

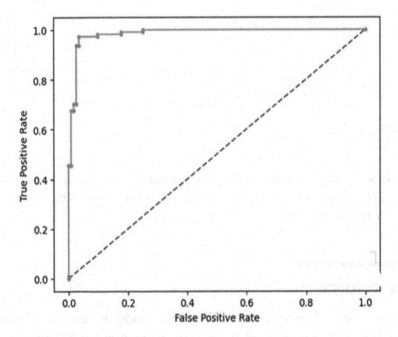

FIGURE 5.7
AUROC curve.

TABLE 5.2

Evaluation Parameters

Algorithm	Classification Accuracy (%)	Precision	Recall	F1 Score
Decision Tree	0.98	0.96	0.98	0.96
Bagging Decision Tree	0.99	0.98	0.98	0.98
Neural Network	0.96	0.97	0.97	0.97
Bagging Neural Network	0.99	0.98	0.98	0.98

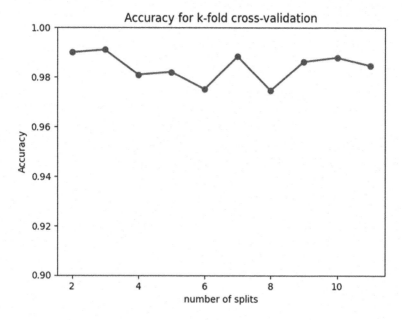

FIGURE 5.8
KFCV.

mean accuracy is generated from the average results, as shown in Table 5.2 [27]. Figure 5.8 depicts the accuracy attained in each iteration of the KFCV technique.

5.4 RESULTS

In the proposed system, the results of the model are explained in reference to assessing parameters. In order to analyze the effectiveness of the system, the KFCV data is also provided [28–32]. The model attained accuracy, sensitivity, and specificity of 0.99, 0.98, and 0.98, respectively, and an AUROC value of 0.998, which is considered among the good results. Another measure to check the authenticity of the model, nine-fold cross-validation, has been executed. The dataset was divided into nine folds, eight of which were employed for training the ML model, and the remaining one fold was set aside for assessing the model.

5.5 CONCLUSION

To date, the yield obtained from crop production is inadequate due to insufficient understanding of the factors influencing production and reliance on conventional farming methods. The recommended system based on ensemble learning will help the agricultural community overcome the corresponding challenges involved. Ultimately, this will contribute to an increase in crop production, leading to higher yields. The proposed model attained sensitivity, specificity, accuracy, and AUROC of 0.98, 0.98, 0.99, and 0.998, respectively. Among the proposed models, the ensemble models outperformed the other two single-classifier models [33–37]. The proposed model will save the efforts of agriculturists required to select the best crop, leading to higher yields automatically. However, it is relevant to indicate that this model is based on finite parameters. However, there are several issues that can be addressed in the future, such as taking other parameters into consideration for model training and tuning other parameters of the proposed models.

REFERENCES

[1] Z. Doshi, S. Nadkarni, R. Agrawal and N. Shah, "AgroConsultant: Intelligent Crop Recommendation System Using Machine Learning Algorithms," *2018 Fourth International Conference on Computing Communication Control and Automation (ICCUBEA)*, Pune, India, 2018, pp. 1–6. doi: 10.1109/ICCUBEA.2018.8697349

[2] *Why is topography important in agriculture*. n.d. Retrieved January 23, 2023, from https://geopard.tech/blog/topography-and-nutrition-content-in-soil-and-yield/

[3] S. Veenadhari, B. Misra and C. Singh, "Machine Learning Approach for Forecasting Crop Yield Based on Climatic Parameters," *2014 International Conference on Computer Communication and Informatics*, Coimbatore, India, 2014, pp. 1–5. doi: 10.1109/ICCCI.2014.6921718

[4] A. Sharma, A. Jain, P. Gupta and V. Chowdary, "Machine Learning Applications for Precision Agriculture: A Comprehensive Review," *IEEE Access*, vol. 9, pp. 4843–4873, 2021. doi: 10.1109/ACCESS.2020.3048415

[5] *2018 Fourth International Conference on Computing Communication Control and Automation (ICCUBEA)*. n.d.

[6] S. M. Pande, P. K. Ramesh, A. Anmol, B. R. Aishwarya, K. Rohilla and K. Shaurya, "Crop Recommender System Using Machine Learning Approach," *2021 5th International Conference on Computing Methodologies and Communication (ICCMC)*, Erode, India, 2021, pp. 1066–1071. doi: 10.1109/ICCMC51019.2021.9418351

[7] R. Kumar, M. P. Singh, P. Kumar and J. P. Singh, "Crop Selection Method to Maximize Crop Yield Rate Using Machine Learning Technique," *2015 International Conference on Smart Technologies and Management for Computing, Communication, Controls, Energy and Materials (ICSTM)*, Avadi, India, 2015, pp. 138–145. doi: 10.1109/ICSTM. 2015.7225403

[8] R. K. Rajak, A. Pawar, M. Pendke, P. Shinde, S. Rathod and A. Devare, "Crop Recommendation System to Maximize Crop Yield using Machine Learning Technique," *International Research Journal of Engineering and Technology*, vol. 4, 2017. Retrieved from www.irjet.net

[9] A. Savla, P. Dhawan, H. Bhadada, N. Israni, A. Mandholia and S. Bhardwaj, "Survey of Classification Algorithms for Formulating Yield Prediction Accuracy in Precision Agriculture," *Innovations in Information, Embedded and Communication systems (ICIIECS)*, 2015.

[10] A. T. M. Shakil Ahamed, N. T. Mahmood, N. Hossain, M. T. Kabir, K. Das, F. Rahman and R. M. Rahman, "Applying Data Mining Techniques to Predict Annual Yield of Major Crops and Recommend Planting Different Crops in Different Districts in Bangladesh," *(SNPD) IEEE/ACIS International*, 2015.

[11] G. Suresh, A. Senthil Kumar, S. Lekashri and R. Manikandan, "Efficient Crop Yield Recommendation System Using Machine Learning For Digital Farming," *International Journal of Modern Agriculture*, vol. 10, no. 1, pp. 906–914, 2021.

[12] Crop Recommendation Dataset|Kaggle. n.d. Retrieved January 23, 2023, from https://www.kaggle.com/datasets/atharvaingle/crop-recommendation-dataset

[13] S. Krishnan, M. J. Franklin, K. Goldberg, J. Wang and E. Wu, "ActiveClean: An Interact Data Cleaning Framework for Modern Machine Learning," *Proceedings of the ACM SIGMOD International Conference on Management of Data*, 26-June-2016, pp. 2117–2120. doi: 10.1145/2882903.2899409

[14] P. Li, X. Rao, J. Blase, Y. Zhang, X. Chu and C. Zhang, "CleanML: A Study for Evaluating the Impact of Data Cleaning on ML Classification Tasks," *Proceedings - International Conference on Data Engineering*, April-2021, pp. 13–24. doi: 10.1109/ ICDE51399.2021.00009

[15] C. M. Salgado, C. Azevedo, H. Proença and S. M. Vieira, "Noise versus Outliers," in MIT Critical Data (ed) *Secondary Analysis of Electronic Health Records*, Springer, Cham (CH), 2016. PMID: 31314268.

[16] I. Izonin, R. Tkachenko, N. Shakhovska, B. Ilchyshyn and K. K. Singh, "A Two-Step Data Normalization Approach for Improving Classification Accuracy in the Medical Diagnosis Domain," *Mathematics*, vol. 10, no. 11, 1942, 2022.

[17] *Why Use Ensemble Learning? - MachineLearningMastery.com*. n.d. Retrieved October 5, 2023, from https://machinelearningmastery.com/why-use-ensemble-learning/

[18] *Ensemble Learning Methods: Bagging, Boosting and Stacking*. n.d. Retrieved October 5, 2023, from https://www.analyticsvidhya.com/blog/2023/01/ensemble-learning-methods-bagging-boosting-and-stacking/

[19] I. D. Mienye and Y. Sun, "A Survey of Ensemble Learning: Concepts, Algorithms, Applications, and Prospects," *IEEE Access*, vol. 10, pp. 99129–99149, 2022.

[20] P. Yadav, *Decision Tree in Machine Learning*. Towards Data Science. n.d. Retrieved March 1, 2023, from https://towardsdatascience.com/decision-tree-in-machine-learning-e380942a4c96

[21] H. H. Patel and P. Prajapati, "Study and Analysis of Decision Tree Based Classification Algorithms," *International Journal of Computer Sciences and Engineering*, vol. 6, no. 10, pp. 74–78, 2018.

[22] P. Picton and P. Picton, *What Is a Neural Network?* Macmillan Education UK, pp. 1–12, 1994.

[23] A. D. Dongare, R. R. Kharde and A. D. Kachare, "Introduction to Artificial Neural Network," *International Journal of Engineering and Innovative Technology (IJEIT)*, vol. 2, no. 1, pp. 189–194, 2012.

[24] J. Madhuri and M. Indiramma, "Artificial Neural Networks Based Integrated Crop Recommendation System Using Soil and Climatic Parameters," *Indian Journal of Science And Technology Conference*, 2021. doi: 10.17485/IJST/v14i19.64

[25] Domino Data Science Dictionary. *What Is Model Evaluation?* n.d. Retrieved March 4, 2023, from https://www.dominodatalab.com/data-science-dictionary/model-evaluation

[26] S. Narkhede, "Understanding AUC-ROC Curve," *Towards Data Science*, vol. 26, no. 1, pp. 220–227, 2018.

[27] *A Gentle Introduction to k-Fold Cross-Validation - MachineLearningMastery.com*. n.d. Retrieved March 4, 2023, from https://machinelearningmastery.com/k-fold-cross-validation/

[28] S. N. H. Bukhari, J. Webber and A. Mehbodniya, "Decision Tree Based Ensemble Machine Learning Model for the Prediction of Zika Virus T-Cell Epitopes as Potential Vaccine Candidates," *Scientific Reports*, vol. 12, 7810, 2022, doi: 10.1038/s41598-022-11731-6

[29] G. S. Raghavendra, S. S. C. Mary, P. B. Acharjee, V. L. Varun, S. N. H. Bukhari, C. Dutta and I. A. Samori, "An Empirical Investigation in Analysing the Critical Factors of Artificial Intelligence in Influencing the Food Processing Industry: A Multivariate Analysis of Variance (MANOVA) Approach," *Journal of Food Quality*, vol. 2022, Article ID 2197717, 7 pages, 2022. doi: 10.1155/2022/2197717

[30] S. N. H. Bukhari, A. Jain, E. Haq, A. Mehbodniya and J. Webber, "Ensemble Machine Learning Model to Predict SARS-CoV-2 T-Cell Epitopes as Potential Vaccine Targets," *Diagnostics*, vol. 11, no. 11, p. 1990, 2021. MDPI AG. doi: 10.3390/diagnostics11111990

[31] S. L. Bangare, D. Virmani, G. R. Karetla, P. Chaudhary, H. Kaur, S. N. H. Bukhari and S. Miah, "Forecasting the Applied Deep Learning Tools in Enhancing Food Quality for Heart Related Diseases Effectively: A Study Using Structural Equation Model Analysis," *Journal of Food Quality*, vol. 2022, Article ID 6987569, 8 pages, 2022. doi: 10.1155/2022/6987569

[32] C. M. Anoruo, S. N. H. Bukhari and O. K. Nwofor, "Modeling and Spatial Characterization of Aerosols at Middle East AERONET Stations," *Theoretical and Applied Climatology*, vol. 152, pp. 617–625, 2023. doi: 10.1007/s00704-023-04384-6

[33] F. Masoodi, M. Quasim, S. Bukhari, S. Dixit and S. Alam, *Applications of Machine Learning and Deep Learning on Biological Data*. CRC Press, 2023.

[34] S. N. H. Bukhari, A. Jain and E. Haq, "A Novel Ensemble Machine Learning Model for Prediction of Zika Virus T-Cell Epitopes," in Gupta, D., Polkowski, Z., Khanna, A., Bhattacharyya, S., and Castillo, O. (eds) *Proceedings of Data Analytics and Management. Lecture Notes on Data Engineering and Communications Technologies*, vol. 91. Springer, Singapore. doi: 10.1007/978-981-16-6285-0_23

[35] S. N. H. Bukhari, F. Masoodi, M. A. Dar, N. I. Wani, A. Sajad and G. Hussain, "Prediction of Erythemato-Squamous Diseases Using Machine Learning," in *Auerbach Publications eBooks*, pp. 87–96, 2023. doi: 10.1201/9781003328780-6

[36] S. N. H. Bukhari, A. Jain, E. Haq, A. Mehbodniya and J. Webber, "Machine Learning Techniques for the Prediction of B-Cell and T-Cell Epitopes as Potential Vaccine Targets with a Specific Focus on SARS-CoV-2 Pathogen: A Review," *Pathogens*, vol. 11, no. 2, p. 146, 2022. MDPI AG. doi: 10.3390/pathogens11020146

[37] S. Nisar, H. Bukhari and M. A. Dar, "Using Random Forest to Predict T-Cell Epitopes of Dengue Virus," *Advances and Applications in Mathematical Sciences*, vol. 20, no. 11, pp. 2543–2547, 2021.

6

A Perusal of Machine-Learning Algorithms in Crop-Yield Predictions

Anshika Gupta, Mohit Soni, and Kalpana Katiyar
Dr. Ambedkar Institute of Technology for Handicapped, Kanpur, India

6.1 INTRODUCTION

Agriculture is crucial in resolving urgent societal issues, particularly in the context of food provision. Currently, a significant section of the world population suffers from hunger, which is mostly caused by food shortages and insufficiencies, which are aggravated by a continually expanding population [1]. The consequences of population increase, natural temperature changes, soil erosion, and changing climate-related needs necessitate novel techniques to ensure constant and timely food cultivation and output. Furthermore, there is an urgent need to promote sustainability in agricultural food production [2]. These requirements highlight the critical need for land evaluation, crop protection, and accurate crop-yield projection in increasing global food production [3]. Crop productivity is closely connected to a slew of factors, including topography, soil quality, insect pressures, genetic characteristics, water availability, weather patterns, and harvest planning [4]. Crop-yield estimation procedures and approaches are both time-specific and fundamentally nonlinear. The integration of a broad variety of interdependent aspects, many of which are controlled by uncontrollable external forces, adds to the complexity [5].

Farmers have traditionally relied on personal experiences and historical data to anticipate crop yields and make important production decisions based on these forecasts. However, in recent years, novel technologies such as crop model simulations and machine learning (ML) have emerged, with the potential for more accurate yield estimates and the capacity to analyze

DOI: 10.1201/9781003485179-6

enormous datasets utilizing high-performance computers. Several research studies are now revealing the greater potential of ML algorithms in this domain as compared to traditional statistical techniques [6].

A subset of artificial intelligence (AI), ML, allows computers to learn without explicit programming. These strategies overcome the limits of both linear and nonlinear agricultural systems, improving predictability. ML models learn by being exposed to certain tasks and patterns in training data. Following that, these models apply their learned information to fresh data, assuming that the knowledge obtained during the training phase is transferrable and relevant. A profusion of instructive and unique research, as well as reviews, have been published in the field of agricultural output assessment over the last 15 years [7] and conducted a thorough assessment of practical remote sensing applications in palm oil farming, finding research gaps and providing viable solutions. It should be noted, however, that their study did not primarily focus on predicting techniques for palm oil production. Researchers [8] investigated contemporary developments in official statistics generation, including remote sensing, surveys, and the integration of multiple data sources, such as meteorological and administrative data. While their research identified the potential to improve current crop production forecasting methods, it also acknowledged the inherent uncertainty involved with such forecasts. Nonetheless, their article, although examining agricultural yield strategies [9], provided a well-structured review that delves into a variety of features and prediction techniques. The focus of their essay, however, was on information extraction rather than identifying research gaps, making recommendations, or critically analyzing previous work. Their research advised a thorough investigation of new parameters for forecasting agricultural production. Another study [10] looked at palm oil productivity based on physiological plant variables, with the goal of providing insights into the causes causing palm oil output gaps. The researchers [11] concentrated on the practical use of ML in agriculture. Within this framework, they examined studies on animal, crop, water, and land management. Another study [12] attempted to determine the ripeness of fruits in order to optimize harvest scheduling and production forecasts.

ML is a subset of AI that focuses on developing efficient approaches for yield prediction based on a wide range of influencing factors. ML extracts important insights from datasets by identifying observable patterns and relationships within the data [13]. These ML systems require training with

| Data | Input the data into the system | Apply logic on the data provided | Get the desired output |

FIGURE 6.1

An ML technique.

datasets that contain outcomes resulting from previous experiences in order to develop prediction models.

The construction of the predictive system requires the use of several characteristics, and the system's parameters are defined using historical data during the training phase [14]. Following that, in the testing phase, a subset of the past data that was not used during training is used to evaluate the system's development. Figure 6.1 depicts an ML technique.

6.2 LITERATURE REVIEW

6.2.1 Crop-Yield Prediction Methods: Traditional vs. Machine Learning

In the past, crop-yield prediction relied more on manual approaches rather than technology-driven methods. These traditional techniques encompassed practices like referencing historical data, monitoring weather conditions, conducting soil analysis, and performing field inspections, among others. However, a notable drawback of these conventional methods was their sluggishness, limited accuracy, and the time they consumed. Numerous sectors employ ML techniques, including supermarkets using ML to forecast customer phone usage and analyze consumer behavior. In the realm of agriculture, ML has been a long-standing tool. One of the complex challenges in precision agriculture revolves around predicting crop production, and various models have been developed and demonstrated their effectiveness in addressing this challenge. Given the vast array of factors influencing agricultural production, such as climate, weather, soil conditions [15], fertilizer application, and seed varieties, tackling this issue necessitates the utilization of multiple datasets. This underscores that the prediction of agricultural yields entails a multifaceted and intricate process, far from being a straightforward task [16]. Present-day crop-yield prediction models exhibit a reasonably accurate ability to anticipate actual yields, although there remains a continuous pursuit for higher prediction performance [17]. We undertook a comprehensive analysis of the existing literature through a systematic literature review (SLR) to gain a comprehensive understanding of the research conducted on the utilization of ML [18] in predicting agricultural production. Such a rigorous SLR study [19] not only brings forth fresh perspectives but also enhances the comprehension of the current state of the art, which is particularly beneficial for newcomers entering the field of research.

6.2.2 Key Challenges in Crop-Yield Prediction

While ML holds significant potential for enhancing crop analysis and prediction, it is essential to account for various challenges that must be addressed.

i. *Data Volume*: A significant amount of data is frequently required for ML models to be trained effectively. Large dataset management and collection might be challenging in the agriculture industry, especially for small farms [20].

ii. *Data Quality*: The caliber of the training data determines how accurate and reliable ML models are. Due to changes in the soil, climate, geography, and other environmental factors, collecting high-quality data in agriculture can be difficult. As a result, gathering and processing data might be labor-intensive. The main difficulties with fruit identification and recognition using deep learning are explored in reference [21]. They found that several issues, including the dearth of high-quality fruit datasets, the detection of small target fruits, fruit identification in obscured and densely populated environments, the detection of fruits with varying sizes and species, and the development of lightweight fruit detection models are to blame for low accuracy, slow processing speed, and limited robustness in fruit detection and recognition.

iii. *Model Complexity*: Given the complexity of agricultural systems, constructing ML models that comprehensively consider all of the significant variables influencing the development and yield of crops can be a daunting task. Among the most often used approaches are neural networks (NN), random forests (RF), support vector machines (SVM), decision trees (DT), and Naive Bayes algorithms.

These algorithms have considerable hurdles, most notably dealing with large datasets, which increase training time complexity and processing needs, especially in the case of SVM. Furthermore, modifying the algorithm for each situation is difficult in the setting of RF [22]. The creation of a solid big data architecture is one of the most difficult challenges, needing both flexibility and great scalability; researchers [23] examined the parameters controlling soil temperature and determined that the link between factors influencing soil temperature is complex and demanding.

iv. *Accessibility*: Securing the required hardware and software infrastructure for the development and deployment of ML models can present difficulties when resources are constrained.

v. *Human Factor*: It may take additional time for farmers and other stakeholders to prepare themselves for adopting novel techniques and technologies, such as ML-based systems. To encourage wider adoption, technology should be designed to be more user-friendly, accessible, and capable of delivering tangible benefits.

6.3 COLLECTION AND PREPROCESSING OF DATA

6.3.1 Data Sources: Remote Sensing, Internet of Things, Weather Stations, and So On

The systematic gathering and measuring of information relating to variables' interests are what data collection includes. This procedure allows researchers to address specific study questions, test hypotheses, and evaluate outcomes. The act of gathering data is a key part of research that cuts across academic fields such as natural and social sciences, humanities, and business. Although the individual methods used vary depending on the field, the overriding aim of guaranteeing precise and accurate data collection remains constant.

i. *Data Collection Using Remote Sensing*
   Evaluating the well-being of crops throughout their growth stages: making predictions about their yield. Remote sensing plays a role in monitoring the health of crops in a specific field on an ongoing basis. Its main purpose is to detect any threats and take measures before

they cause any harm. Additionally [11], it serves as a method for estimating the yield of the cultivation cycle and determining the time for harvesting based on weather conditions. CropIn's SmartRisk is an example of a product that utilizes satellite-based sensing technology combined with processing capabilities to provide accurate predictive insights about the entire cultivated area. Remote sensing has the enormous advantage of allowing growers to continually monitor their fields independent of their physical location. This is especially beneficial for large agribusinesses with farms scattered across the country since all farms must produce products of consistent quality [24]. Remote sensing allows all stakeholders to work together to follow the crop's progress through various phases, ensuring that it satisfies the business's requirements and maintains uniform quality. Crop management software, such as CropIn's SmartFarm, uses remote sensing technologies to give detailed insights about crop development and performance throughout the growing season. Researchers [25] provide further information about remote sensing and its application in agriculture.

ii. *Data Collection Using the Internet of Things*

Collecting Internet of Things (IoT) data involves the utilization of sensors to oversee the operation of IoT-connected devices. These sensors continuously observe the status of the IoT network, gathering and transmitting real-time data that can be stored and retrieved whenever necessary. There are different types of IoT data. Different types of IoT data are mentioned next:

- *Submeter Data*: Property owners can utilize submetering to measure individual utility use in multi-user environments. Submeters can collect data in buildings with several tenants using water, electricity, gas, or cable resources.
- *Environmental Data*: IoT sensors can measure and monitor environmental data such as humidity, temperature, movement, and air quality.
- *Equipment Data*: It includes the data that pertains to the state of IoT devices. Equipment data is collected in real time to allow predictive maintenance tasks.
- Current uses provide insights into the effects of the IoT and new practices. Nonetheless, with the advancement of technology [26], IoT technologies may play a critical part in numerous farming tasks. These include using communication infrastructure, data

acquisition, smart objects and sensors, mobile devices, exploiting cloud-based intelligent information, easing decision-making, and automating agricultural activities. IoT technology oversees the remote monitoring of plants and animals while collecting data from mobile phones and gadgets [27]. Researchers have put forth diverse approaches, architectural solutions, and a range of equipment to monitor and transmit crop-related information across different growth stages tailored to various crop types and field conditions. Numerous manufacturers offer communication devices, multiple sensors, robotic systems, heavy machinery, and drones designed for data collection and dissemination purposes. Additionally, food and agriculture organizations, in collaboration with government bodies, are actively formulating guidelines and policies to regulate the use of these technologies, intending to ensure food safety and environmental preservation [28].

iii. *Data Collection Using the Weather Stations*

Weather stations are buildings or objects with a variety of equipment and sensors that gather and measure meteorological information and local environmental variables. These stations are positioned carefully to keep track of variables, including temperature, humidity, atmospheric pressure, wind direction, speed, precipitation (rain and snowfall), and, occasionally, even more parameters like solar radiation and soil moisture. Simple home weather stations for personal use can range in complexity from highly sophisticated automated systems employed by meteorological authorities and research institutions. Weather data is frequently made publicly available for use in a variety of applications and decision-making processes by modern weather stations that frequently send data in real-time to centralized databases.

Farmers need to learn how to monitor environmental conditions properly and effectively, especially for plants that are particularly sensitive to the weather, to reduce the effects of climate change on diverse crops. Two popular techniques for gathering information and monitoring weather conditions are on-site sensors and weather stations. Although sensors can gather precise meteorological data locally, installing and maintaining them may be expensive and time-consuming. Using online weather stations, which are often controlled by the government and accessible to everyone, is an option; nevertheless, their accuracy is debatable because of how frequently they are situated far [29] from farmers' greenhouses. As a result, we contrasted the precision of kriging estimators utilizing data from local sensors and weather station data gathered by the Central Weather Bureau. Temperature data were interpolated using the spatiotemporal kriging technique. For comparison, the actual value in the greenhouse's center was utilized. Our findings showed that the local sensor estimator had somewhat higher accuracy than the weather station estimator. On-site sensors can provide farmers with precise estimates of environmental data, but if they are not accessible, utilizing a nearby weather station estimation is also acceptable (Figure 6.2) [30].

6.3.2 Cleaning, Transformation, and Integration of Data

Preparing raw data for analysis and model training by cleaning and preprocessing it are crucial phases in the ML pipeline. These stages are essential since your data's applicability and quality have a big impact on the effectiveness and precision of your ML models.

i. **Data Cleaning**: Finding and fixing flaws or inconsistencies in your dataset is known as data cleaning. It strives to guarantee the accuracy and dependability of your data. Data cleansing activities frequently involve the following:

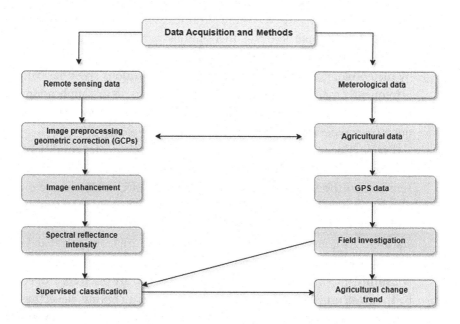

FIGURE 6.2

Various methods of data collection.

- *Handling Missing Values*: Removing rows with missing values, imputing missing values with a default value or a statistical measure (such as the mean or median), or using more sophisticated imputation techniques are all examples of how to handle missing values.
- *Outlier Detection and Treatment*: Determining whether to eliminate, transform, or leave as-is outliers (data points notably different from the majority) based on the context.
- *Data Formatting*: Data formatting involves making sure that the data types are acceptable and consistent, converting them if required, and formatting the data according to the needs of the ML algorithm.
- *Handling Duplicates*: Detecting and getting rid of duplicate entries will prevent bias and redundancy in the dataset.
- *Noise Reduction*: Data smoothing or filtering techniques are used to reduce noise, especially in time-series or sensor data. Researchers [31] use big data with random sample blocks to explain the process of data cleaning and preprocessing.

ii. **Data Preprocessing**: The translation of cleansed data into a format appropriate for ML algorithms is known as data preparation. Researchers [32] explain the process of data preprocessing combinations to improve the performance of quality classification in the manufacturing process. Several tasks are included in this process, including the following:

- *Feature Scaling*: Feature scaling is the process of reducing numerical characteristics to a common range (e.g., normalization or standardization) to prevent features from dominating others during model training.
- *Feature Engineering*: This is the process of developing new features or modifying existing ones to extract useful information or enhance model performance. This includes the one-time encoding of categorical variables, the creation of interaction features, and the extraction of date/time features.
- *Data Encoding*: Transforming variables that are categorical into numerical format using techniques such as one-hot encoding or label encoding.
- *Normalization*: Normalization is the process of scaling features to have a zero mean and unit variance so that ML algorithms can converge faster [33] (Figure 6.3).

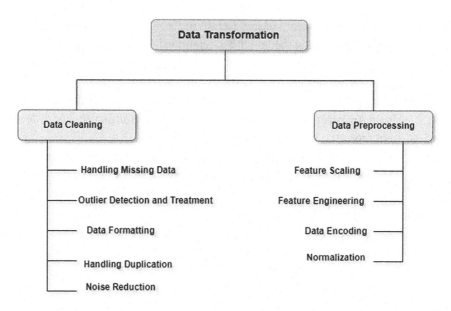

FIGURE 6.3

Methods of data cleaning and preprocessing.

6.3.3 Crop-Yield Prediction Using Feature Engineering

Feature design or engineering is an important stage in preparing data for ML algorithms. It entails generating new features (also known as variables or attributes) from current data to improve the performance of an ML model [34]. The purpose of feature engineering is to extract useful information from raw data and represent it in a manner that the ML algorithm can learn from it successfully. Feature engineering helps us to extract the dependent and independent variable that fits the ML model and helps us in the further fine-tuning of preconceived beliefs and probability. Researchers [35] use the weather data and the historical data of maize and the Irish potato for future prediction of the yield of the Irish potato and maize (Figure 6.4).

6.4 MACHINE-LEARNING ALGORITHMS FOR CROP-YIELD PREDICTION

ML algorithms are software programs that can detect hidden patterns in data, make predictions, and improve their performance by self-learning

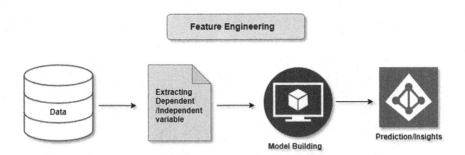

FIGURE 6.4

Feature engineering.

from experience. ML employs a variety of algorithms, each adapted to a certain purpose [36]. Simple linear regression, for example, is well-suited for prediction tasks such as anticipating stock market movements, but the K-nearest neighbors (KNN) method is well-suited for classification challenges. The ML tasks are majorly divided into classification and regression tasks.

a. **Classification Task**: Classification tasks are the tasks that necessitate the application of ML algorithms to learn how to assign a class label to problem domain instances [37]. One simple example is categorizing emails as "spam" or "not spam." The most popular classification algorithms are logistic regression, Naive Bayes, KNN, and SVM.

b. **Regression Task**: Regression represents a supervised ML approach for forecasting continuous values. The regression algorithm's ultimate purpose is to plot the best-fit line or curve between the data. Variance, bias, and error are the three major metrics used to evaluate the trained regression model [38]. The most popular regression algorithms are Linear regression, Ridge regression, Lasso regression, Polynomial regression, and Bayesian linear regression. More broadly, ML is divided into the following categories: Supervised learning, Unsupervised learning algorithms, and Reinforcement learning algorithms (Figure 6.5).

6.4.1 Linear Regression Models

The linear regression technique is one of the most basic and widely used ML methods. This is a statistical approach for performing predictive analysis. Linear regression forecasts continuous/real or quantitative variables

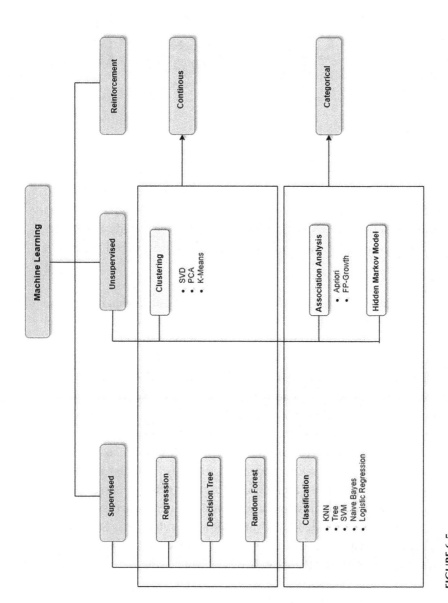

FIGURE 6.5

Different types of ML algorithms.

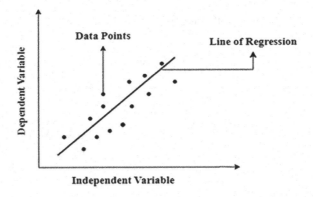

FIGURE 6.6

Linear regression

such as sales, salary, age, product price, and so on. The linear regression technique demonstrates a linear connection between a dependent ($\hat{y}$) variable and one or more independent (X_1) variables, thus the name. Because linear regression demonstrates a linear connection, it determines how the value of the dependent variable evolves concerning the value of the independent variable [38]. The linear regression model represents the association between variables using a slanted straight line. Mathematically linear regression can be represented as: $\hat{y} = b_0 + b_1X_1$, where $\hat{y}$ = Dependent Variable; b_0 = Y – Intercept; b_1 = Slope Coefficient; X_1 = Independent Variable (Figure 6.6).

6.4.2 Decision Tree and Random Forest Models

Decision Tree: A DT is a method of supervised learning that may be used to solve classification and regression problems, while it is mostly used to solve classification problems. It works as a treelike classifier, with core nodes representing dataset attributes, branches representing decision rules, and each leaf node representing a result (Rokach & Maimon, n.d.). There are two types of nodes in a DT: decision nodes and leaf nodes. Decision nodes are used to make decisions and have several branches, whereas leaf nodes indicate the results of those decisions and do not branch out any further. Inside a DT, the algorithm initiates its process at the root node, progressing toward predicting the class for a provided dataset. This procedure involves examining the root node's property values against the attributes of the actual dataset record, determining the appropriate branch to follow, and proceeding to the next node based on the outcome of this comparison.

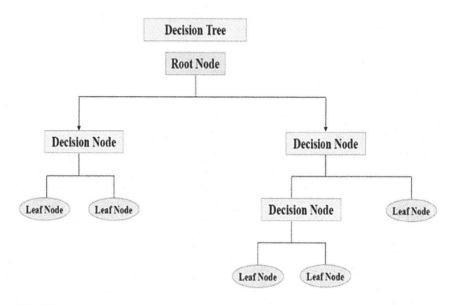

FIGURE 6.7

Decision tree.

The CART algorithm, which stands for classification and regression tree algorithm, is used to form a tree (Figure 6.7).

Random Forest Model: RF is a well-known ML algorithm that is firmly rooted in the realm of supervised learning. It is useful in the field of ML for both classification and regression applications. This approach is based on the notion of ensemble learning, which is a strategy that includes combining numerous classifiers to solve complex issues and improve the model's performance. An "RF" classifier, as the name implies, is a classifier composed of several DTs constructed on varied subsets of the supplied dataset, and it utilizes their aggregate predictive potential to improve dataset accuracy. Rather than depending exclusively on a single DT, the RF aggregates forecasts from each tree and then creates a final prediction based on a majority vote of these projections. The presence of a higher number of trees inside the forest adds to improved accuracy and protects against overfitting. Because the RF mixes numerous trees to forecast the dataset's class/output, some DTs may predict the proper output while others may not (Figure 6.8).

6.4.3 Support Vector Machine

The SVM is a supervised ML algorithm that may be used for classification as well as regression tasks. While SVM can handle regression issues,

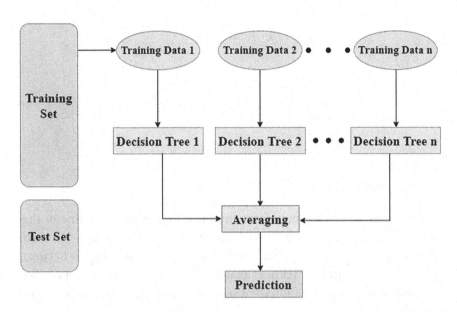

FIGURE 6.8
RF classifier.

it is best suited for classification jobs. Its major purpose is to find the best hyperplane inside an N-dimensional space capable of properly classifying data points in the feature space. The hyperplane's size is determined by the amount of input characteristics and is intended to maximize the gap between the nearest points of separate classes. When the number of features exceeds three, it becomes difficult to visualize the hyperplane. The SVM is a robust ML technique used for applications such as linear or nonlinear classification, regression, and even outlier identification. Text classification, picture classification, spam identification, handwriting recognition, gene expression analysis, face detection, and anomaly detection are just a few of the tasks that SVMs are used for. Because of their capacity to handle high-dimensional data and account for nonlinear interactions, their versatility and efficiency show through a wide range of applications. When attempting to discover the largest separating hyperplane among several classes within the target feature, the efficiency of SVM algorithms is clear. Let's consider two independent variables Y_1 and Y_2, and one dependent variable, which is either a green circle or a yellow circle.

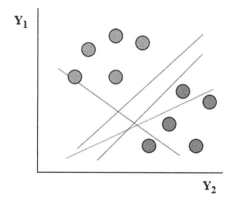

FIGURE 6.9

Linearly separable data points.

As seen in Figure 6.9, numerous lines (our hyperplane is a line in this case because we only have two input characteristics, Y_1 and Y_2) exist, thereby splitting our data points and conducting a classification between the red and blue circles. Mekni et al. uses an SVM algorithm as supervised learning for the prioritization of novel potential of the SARS-CoV-2 main protease inhibitor and explains how an SVM [39].

6.4.4 Neural Network Model

An NN is an AI technology that allows computers to interpret data like that of the human brain. It is classified under ML, especially a branch known as deep learning. This method organizes linked nodes or neurons into layered structures resembling the human brain. This design creates an adaptive system in which computers may learn from mistakes and gradually improve their performance. As a result, artificial NNs attempt to perform complicated tasks with greater precision, such as document summarization or facial recognition. There are various applications of NN, a few of them are listed next:

Natural Language Processing: Natural language processing (NLP) is the ability to analyze human-generated text in a manner that mimics human language comprehension. For computers to extract insights and comprehension from textual data and texts, NNs play a critical role. NLP has a variety of practical applications, such as the following:
- Giving automated virtual agents and chatbots more power.
- Organizing and categorizing textual stuff automatically.

- Enabling business intelligence analysis of large documents like emails and forms.
- Indexing key terms that express attitude, such as good and negative social media comments.
- Creating document summaries and essays on certain subjects.

Computer Vision: Computer vision refers to a computer's ability to extract information and comprehend visual content from images and videos. Using NN, computers are now capable of distinguishing and classifying pictures in the same way that people do. Computer vision has practical applications in a variety of fields, including the following:

- Visual recognition in self-driving vehicles allows them to recognize road signs and other vehicles on the road
- Content moderation, which involves automating the identification and removal of potentially hazardous or improper content from picture and video collections
- Facial recognition for recognizing certain facial characteristics such as open eyes, spectacles, or facial hair
- Image labeling for features, such as company logos, clothes, safety equipment, and other image information

Convolutional Neural Networks (CNN): CNNs are a type of multi-layered NN that is specially designed for processing two-dimensional input, such as pictures and videos. CNNs are inspired by earlier work on time-delay neural networks (TDNN), which optimize learning calculations by weight sharing in a temporal dimension and are particularly suited for speech and time-series data processing applications [40]. CNNs were a significant milestone in the field of deep learning by effectively robustly training deep hierarchies of layers. These networks use an architectural approach that capitalizes on spatial and temporal correlations, minimizing the number of parameters that need to be learned and resulting in better performance than traditional feed-forward back-propagation training [41].

CNNs are distinguished by their low data preprocessing needs. Images are broken into tiny parts in CNNs, which serve as inputs to the lowest layer of the hierarchical structure. Information passes via several network levels, with each layer using digital filtering techniques to extract important data properties. This method gives the network some resistance to shifts, scaling, and rotations, owing to the local receptive field, which gives neurons or processing units access to basic properties, such as aligned edges and corners.

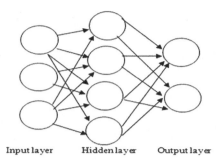

Input layer   Hidden layer   Output layer

FIGURE 6.10
An ANN model.

> *Artificial Neural Network (ANN)*: An ANN is a network of linked nodes that is remote, like the huge network of neurons in the brain, as seen in Figure 6.10. Each circular node represents an artificial neuron [42], and each arrow symbolizes a link from one neuron's output to the input of another, which should (hopefully) be capable of handling this. An ANN is made up of three types of layers: input, hidden, and output. The hidden layer connects the input and output layers.

6.4.5 Ensemble Models

Ensemble learning is a comprehensive meta-strategy in ML that aims to improve predicted accuracy by combining predictions from many models. While there appears to be an infinite number of ensembles that may be built for predictive modeling, three approaches dominate the field of ensemble learning. In fact, these procedures have grown into entire disciplines of study, giving rise to a plethora of specialized techniques within each. Bagging, stacking, and boosting are the three primary classes of ensemble learning methods, and it is critical to understand each approach and include it in the predictive modeling project.

6.5 CASE STUDIES

6.5.1 Prediction of Crop Yield for a Specific Crop Type

Shahram Rezapour [18] and the rest of his team completed a study that focused on using ML approaches to predict agricultural productivity in arid and semi-arid regions. Their major goal was to address food security challenges, particularly related to climate change's impact on food production. The case study in the issue is about estimating chickpea output in Kermanshah Province [43], which is located in Iran. The researchers used remote sensing data and meteorological information from synoptic stations to build prediction models as important inputs in their technique. They used ML techniques such as support vector regression (SVR) [44], RF [45], and extreme randomized trees (ERT) to analyze the data and provide exact estimates of rainfed agricultural productivity.

The research findings demonstrated that, when evaluating the optimum combination (OC) scenario, the RF strategy produced the lowest errors among the numerous approaches studied. Furthermore, in this situation, the RF technique indicated surface moisture, minimum air humidity, and subsurface humidity as the most important parameters for accurate forecasts in March, January, and January, respectively. Furthermore, the researchers suggested that future studies look at different crop kinds, prediction systems, and geographic areas other than chickpeas, broadening the scope to include other agricultural goods. In conclusion, Rezapour and his colleagues' work demonstrates the effectiveness of ML algorithms for predicting rainfed agricultural outputs. Their study provides useful insights into predicting chickpea output in Kermanshah Province, Iran, while emphasizing the need for accurate data collection and analysis in making well-informed agricultural decisions [46]. These findings contribute significantly to ongoing programs targeted at ensuring food security in arid and semi-arid countries.

6.5.2 Dealing with Extreme Weather Events

[47] his study is about assessing and forecasting drought conditions in Jaisalmer, India, using data-driven methodologies. It uses ML techniques, notably the random subspace (RSS) approach, to calculate the standardized precipitation index (SPI) over different periods. The data illustrate the RSS-M5P model's usefulness in capturing monthly SPI trends and providing accurate SPI predictions at several scales, including 3, 6, and 9 months.

The findings of this study have substantial practical implications for a variety of industries, including agriculture, hydrology, and meteorology. Drought prediction models produced by researchers are useful for activities such as optimizing irrigation schedules, modeling crop behavior, managing water resources, running reservoirs, and improving weather forecasts. It emphasizes the importance of developing comprehensive drought management policies. To summarize, this case study focuses on the use of data-driven models to analyze and predict drought in Jaisalmer, India [48]. It calculates the SPI across multiple temporal scales using ML and the RSS approach. These models have practical implications in a variety of industries and can aid in the development of climate-smart farming practices.

6.6 DISCUSSION

The use of ML algorithms in crop production prediction has far-reaching consequences for the agricultural environment. We unleash the potential for more exact and timely production projections by harnessing these advanced computational methods, providing farmers and stakeholders with essential information about the performance of their crops. This results in better decision-making, resource optimization, and risk reduction. Furthermore, improved forecasting ability owing to weather, pests, or illnesses helps to increase food security and economic stability [49]. As ML models improve their forecast accuracy, the agricultural industry will become more resilient and sustainable, influencing global food production and assuring a more secure food supply for our rising population.

Several obstacles and limits must be considered while traversing the area of ML algorithms for agricultural yield projections. One of the most significant issues is data quality and availability [50], as successful projections rely largely on complete datasets that include parameters like weather, soil conditions, and historical yields. Furthermore, model interpretability remains a challenge, prompting research efforts to make these algorithms more visible and accessible for farmers and stakeholders. Another issue to consider is scalability, which is especially important in resource-constrained environments. To address these issues, future research should focus on multi-source data integration, real-time monitoring systems that can adapt to changing conditions, and ML solutions that are tailored for low-resource contexts. ML in crop-yield projections has practical applications in precision agriculture, where farmers may optimize resource allocation, insurance and risk assessment, financial planning, and supply chain management, improving logistics and minimizing waste. As this sector evolves, it has the potential to revolutionize agricultural practices and greatly contribute to global food security. Future research should focus on the interface of ML and climate change adaptation, allowing agriculture to adjust to changing weather patterns and environmental circumstances. More interpretable AI models will boost confidence and usefulness, while ongoing developments in data-gathering methods, sensor technology, and precision agricultural tools will be critical. Finally, these activities will define agriculture's future by promoting sustainable practices and maintaining food security in an ever-changing world.

6.7 CONCLUSION

The agricultural sector must adopt new technology in order to satisfy the demands of a rising global population. Furthermore, timely advice for crop output forecasts is critical for farmers in developing successful production plans. ML frameworks provide significant insights by processing large datasets and analyzing the resulting data. These tools make it easier to create models that represent the interactions between numerous elements and activities. Furthermore, they allow for the prediction of future events in certain contexts. The current study focuses on a wide range of criteria used in the selected studies, with a particular emphasis on data availability and research scope. The majority of the cited works investigate agricultural production predictions using ML algorithms; however, they differ in terms of the vast variety of features used. These investigations also revealed differences in crop kinds, geographic areas, and intensity levels. The characteristics used are determined by the dataset's accessibility and the study aims. While it is difficult to determine the ultimate best technique based on present data, the widespread use of some ML algorithms and their promising performance is remarkable. Specifically, traditional ML architectures like linear regression, RF, and NN show potential. Some deep learning models, such as deep neural networks (DNN), CNN, and long short-term memory (LSTM), are used in crop-yield estimates in addition to these techniques. Future studies should explore feature selection algorithms in combination with existing high-performing models to get a clear conclusion about the best-performing model.

REFERENCES

[1] D. Elavarasan and P. M. D. Vincent, "Crop yield prediction using deep reinforcement learning model for sustainable agrarian applications," *IEEE Access*, vol. 8, pp. 86886–86901, 2020.

[2] J. Huang et al., "Assimilation of remote sensing into crop growth models: Current status and perspectives," *Agric. For. Meteorol.*, vol. 276, p. 107609, 2019.

[3] S. Li, S. Peng, W. Chen, and X. Lu, "INCOME: Practical land monitoring in precision agriculture with sensor networks," *Comput. Commun.*, vol. 36, no. 4, pp. 459–467, 2013.

[4] A. Singh, B. Ganapathysubramanian, A. K. Singh, and S. Sarkar, "Machine learning for high-throughput stress phenotyping in plants," *Trends Plant Sci.*, vol. 21, no. 2, pp. 110–124, 2016.

[5] A. Chlingaryan, S. Sukkarieh, and B. Whelan, "Machine learning approaches for crop yield prediction and nitrogen status estimation in precision agriculture: A review," *Comput. Electron. Agric.*, vol. 151, pp. 61–69, 2018.

[6] D. Elavarasan, D. R. Vincent, V. Sharma, A. Y. Zomaya, and K. Srinivasan, "Forecasting yield by integrating agrarian factors and machine learning models: A survey," *Comput. Electron. Agric.*, vol. 155, pp. 257–282, 2018.

[7] K. L. Chong, K. D. Kanniah, C. Pohl, and K. P. Tan, "A review of remote sensing applications for oil palm studies," *Geo-Spatial Inf. Sci.*, vol. 20, no. 2, pp. 184–200, 2017.

[8] L. J. Young, "Agricultural crop forecasting for large geographical areas," *Annu. Rev. Stat. its Appl.*, vol. 6, pp. 173–196, 2019.

[9] T. Van Klompenburg, A. Kassahun, and C. Catal, "Crop yield prediction using machine learning: A systematic literature review," *Comput. Electron. Agric.*, vol. 177, p. 105709, 2020.

[10] L. S. Woittiez, M. T. Van Wijk, M. Slingerland, M. Van Noordwijk, and K. E. Giller, "Yield gaps in oil palm: A quantitative review of contributing factors," *Eur. J. Agron.*, vol. 83, pp. 57–77, 2017.

[11] K. G. Liakos, P. Busato, D. Moshou, S. Pearson, and D. Bochtis, "Machine learning in agriculture: A review," *Sensors*, vol. 18, no. 8, p. 2674, 2018.

[12] B. Li, J. Lecourt, and G. Bishop, "Advances in non-destructive early assessment of fruit ripeness towards defining optimal time of harvest and yield prediction—A review," *Plants*, vol. 7, no. 1, p. 3, 2018.

[13] G. Nemomsa, "Comparative analytics and predictive modeling of student performance through data mining techniques," *IUP J. Comput. Sci. Forthcom.*, 2018. doi: 10.2139/ssrn.3382511

[14] S. Venkatakrishnan, A. Kaushik, and J. K. Verma, "Sentiment analysis on Google Play Store data using deep learning," in Johri, P., Verma, J., and Paul, S. (eds) *Applications of Machine Learning. Algorithms for Intelligent Systems.* Springer, Singapore, 2020. doi: 10.1007/978-981-15-3357-0_2

[15] X. Wang, W. Hu, K. Li, L. Song, and L. Song, "Modeling of soft sensor based on DBN-ELM and its application in measurement of nutrient solution composition for soilless culture," in *2018 IEEE International Conference of Safety Produce Informatization (IICSPI)*, 2018, pp. 93–97.

[16] S. K. Sharma, D. P. Sharma, and J. K. Verma, "Study on machine-learning algorithms in crop yield predictions specific to indian agricultural contexts," *2021 Int. Conf. Comput. Perform. Eval. ComPE 2021*, no. April 2022, pp. 155–166, 2021. doi: 10.1109/ComPE53109.2021.9752260

[17] P. Muruganantham, S. Wibowo, S. Grandhi, N. H. Samrat, and N. Islam, "A systematic literature review on crop yield prediction with deep learning and remote sensing," *Remote Sens.*, vol. 14, no. 9, p. 1990, 2022.

[18] S. Rezapour et al., "Forecasting rainfed agricultural production in arid and semi-arid lands using learning machine methods: A case study," *Sustainability*, vol. 13, no. 9, 2021. doi: 10.3390/su13094607

[19] M. Chinlampianga, "Traditional knowledge, weather prediction and bioindicators: A case study in Mizoram, Northeastern India," *Indian J. Tradit. Knowl.*, vol. 10, no. 1, pp. 207–211, 2011.

[20] A. Cravero, S. Pardo, S. Sepúlveda, and L. Muñoz, "Challenges to use machine learning in agricultural big data: A systematic literature review," *Agronomy*, vol. 12, no. 3, p. 748, 2022.

[21] F. Xiao, H. Wang, Y. Xu, and R. Zhang, "Fruit detection and recognition based on deep learning for automatic harvesting: An overview and review," *Agronomy*, vol. 13, no. 6, p. 1625, 2023.

[22] S. Del Río, V. López, J. M. Benítez, and F. Herrera, "On the use of mapreduce for imbalanced big data using random forest," *Inf. Sci. (Ny).*, vol. 285, pp. 112–137, 2014.

[23] C. A. Salma, B. Tekinerdogan, and I. N. Athanasiadis, "Domain-driven design of big data systems based on a reference architecture," in Ivan Mistrik (ed) *Software Architecture for Big Data and the Cloud*. Elsevier, 2017, pp. 49–68.

[24] P. Johri, J. K. Verma, and S. Paul, *Applications of Machine Learning*. Springer, 2020.

[25] M. J. M. Cheema, T. Iqbal, A. Daccache, S. Hussain, and M. Awais, "Precision agriculture technologies: Present adoption and future strategies," in *Precision Agriculture*. Elsevier, 2023, pp. 231–250.

[26] J. Sun, W. Gan, H.-C. Chao, and P. S. Yu, "Metaverse: Survey, applications, security, and opportunities," vol. 1, no. 1, pp. 1–35, 2022, [Online]. Available: http://arxiv.org/abs/2210.07990

[27] S. Agarwal and S. Tarar, "A hybrid approach for crop yield prediction using machine learning and deep learning algorithms," *J. Phys. Conf. Ser.*, vol. 1714, no. 1, p. 12012, 2021.

[28] M. Dhanaraju, P. Chenniappan, K. Ramalingam, S. Pazhanivelan, and R. Kaliaperumal, "Smart farming: Internet of Things (IoT)-based sustainable agriculture," *Agriculture*, vol. 12, no. 10, p. 1745, 2022.

[29] A. Haghverdi, R. A. Washington-Allen, and B. G. Leib, "Prediction of cotton lint yield from phenology of crop indices using artificial neural networks," *Comput. Electron. Agric.*, vol. 152, pp. 186–197, 2018.

[30] P.-F. Kuo, T.-E. Huang, and I. G. B. Putra, "Comparing kriging estimators using weather station data and local greenhouse sensors," *Sensors*, vol. 21, no. 5, p. 1853, 2021.

[31] S. Salloum, J. Z. Huang, and Y. He, "Exploring and cleaning big data with random sample data blocks," *J. Big Data*, vol. 6, pp. 1–28, 2019.

[32] E. Cho, T.-W. Chang, and G. Hwang, "Data preprocessing combination to improve the performance of quality classification in the manufacturing process," *Electronics*, vol. 11, no. 3, p. 477, 2022.

[33] P. Nevavuori, N. Narra, and T. Lipping, "Crop yield prediction with deep convolutional neural networks," *Comput. Electron. Agric.*, vol. 163, p. 104859, 2019.

[34] M. Rashid, B. S. Bari, Y. Yusup, M. A. Kamaruddin, and N. Khan, "A comprehensive review of crop yield prediction using machine learning approaches with special

emphasis on palm oil yield prediction," *IEEE Access*, vol. 9, pp. 63406–63439, 2021. doi: 10.1109/ACCESS.2021.3075159

[35] M. Kuradusenge et al., "Crop yield prediction using machine learning models: Case of Irish potato and maize," *Agriculture*, vol. 13, no. 1, p. 225, 2023.

[36] A. K. Tiwari, "Introduction to machine learning," in Arvind Tiwari and Pradeep Kumar (eds) *Ubiquitous Machine Learning and Its Applications*. IGI Global, 2017, pp. 1–14.

[37] M. Yuvalı, B. Yaman, and Ö. Tosun, "Classification comparison of machine learning algorithms using two independent CAD datasets," *Mathematics*, vol. 10, no. 3, p. 311, 2022.

[38] S.-J. Kim, S.-J. Bae, and M.-W. Jang, "Linear regression machine learning algorithms for estimating reference evapotranspiration using limited climate data," *Sustainability*, vol. 14, no. 18, p. 11674, 2022.

[39] N. Mekni, C. Coronnello, T. Langer, M. De Rosa, and U. Perricone, "Support vector machine as a supervised learning for the prioritization of novel potential sars-cov-2 main protease inhibitors," *Int. J. Mol. Sci.*, vol. 22, no. 14, p. 7714, 2021.

[40] F. Nurahmadi, F. Lubis, and P. I. Nainggolan, "Analysis of deep learning architecture in classifying SNI masks," *J. Informatics Telecommun. Eng.*, vol. 5, no. 2, pp. 473–482, 2022.

[41] I. L. Rahmatullah, "Pengenalan suara menggunakan algoritma convolutional neural network pada gim pembelajaran bahasa arab." Fakultas Sains dan Teknologi UIN Syarif Hidayatullah Jakarta.

[42] B. K. Zaied et al., "Prediction and optimization of biogas production from POME co-digestion in solar bioreactor using artificial neural network coupled with particle swarm optimization (ANN-PSO)," *Biomass Convers. Biorefinery*, pp. 1–16, 2020. doi: 10.1007/s13399-020-01057-6

[43] A. Mostafaeipour and E. Jooyandeh, "Prioritizing the locations for hydrogen production using a hybrid wind-solar system: A case study," *Adv. Energy Res.*, vol. 5, no. 2, p. 107, 2017.

[44] M. Yousefi et al., "RETRACTED ARTICLE: Support vector regression methodology for prediction of output energy in rice production," *Stoch. Environ. Res. Risk Assess.*, vol. 29, pp. 2115–2126, 2015.

[45] V. Rodriguez-Galiano, M. Sanchez-Castillo, M. Chica-Olmo, and M. Chica-Rivas, "Machine learning predictive models for mineral prospectivity: An evaluation of neural networks, random forest, regression trees and support vector machines," *Ore Geol. Rev.*, vol. 71, pp. 804–818, 2015.

[46] A. Sharifi, "Yield prediction with machine learning algorithms and satellite images," *J. Sci. Food Agric.*, vol. 101, no. 3, pp. 891–896, 2021.

[47] A. Elbeltagi et al., "Drought indicator analysis and forecasting using data driven models: Case study in Jaisalmer, India," *Stoch. Environ. Res. Risk Assess.*, vol. 37, no. 1, pp. 113–131, 2023. doi: 10.1007/s00477-022-02277-0

[48] B. T. Pham, I. Prakash, and D. T. Bui, "Spatial prediction of landslides using a hybrid machine learning approach based on random subspace and classification and regression trees," *Geomorphology*, vol. 303, pp. 256–270, 2018.

[49] K. B. Banakara, N. Sharma, S. Sahoo, S. K. Dubey, and V. M. Chowdary, "Evaluation of weather parameter-based pre-harvest yield forecast models for wheat crop: A case study in Saurashtra region of Gujarat," *Environ. Monit. Assess.*, vol. 195, no. 1, 2023. doi: 10.1007/s10661-022-10552-4

[50] N. Kim, K. J. Ha, N. W. Park, J. Cho, S. Hong, and Y. W. Lee, "A comparison between major artificial intelligence models for crop yield prediction: Case study of the midwestern United States, 2006–2015," *ISPRS Int. J. Geo-Information*, vol. 8, no. 5, 2019. doi: 10.3390/ijgi8050240

7

Harvesting Intelligence: AI and ML Revolutionizing Agriculture

Arya Kumari, Muhammad Najeeb Khan,
and Amit Kumar Sinha
Shri Mata Vaishno Devi University, Katra, India

7.1 INTRODUCTION

Agriculture has been the backbone of human civilization for millennia, and with the advent of technology, it is undergoing a remarkable transformation. Machine learning (ML), with its ability to process vast amounts of data and extract meaningful insights, is playing a pivotal role in revolutionizing the agricultural industry. In this chapter, we will explore the diverse applications of ML in agriculture and how it is enhancing productivity, sustainability, and overall efficiency.

7.2 PRECISION FARMING AND CROP MANAGEMENT

7.2.1 Crop Yield Prediction

ML plays a crucial role in crop yield prediction, as it can leverage historical and real-time data to build predictive models that estimate future crop yields accurately. Here's the pictorial representation (Figure 7.1) of how ML can be used in crop yield prediction:

Now let's see a brief explanation of each step, as shown in Figure 7.1

1. Data Collection and Preprocessing:
   - Clean and preprocess the data to handle missing values and outliers, and ensure data consistency.

DOI: 10.1201/9781003485179-7

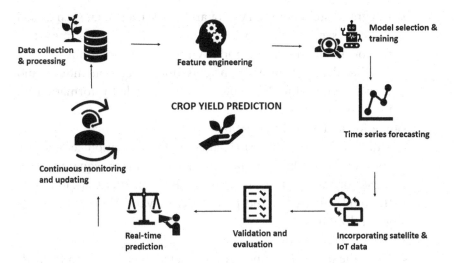

FIGURE 7.1
Crop yield prediction with the help of ML.

2. Feature Engineering:
   - Select and engineer relevant features from the collected data that can influence crop yields, such as temperature, precipitation, soil moisture, and fertilization methods.
3. Model Selection:
   - By choosing an appropriate ML algorithm based on the nature of the problem (some commonly used algorithms are linear regression, decision trees, random forests, support vector machines (SVM), neutral network, time series forecasting, and many more) and the size of the dataset, we can predict crop yield.
4. Model Training:
   - By splitting the dataset into training and validation sets, we evaluate the model's performance and train the chosen ML model on the training set, adjusting the models parameters to minimize prediction errors.
5. Time Series Forecasting:
   - For crop yield prediction involving time series data, select models specifically designed for time-dependent patterns, such as autoregressive integrated moving average (ARIMA), seasonal decomposition of time series (STL), or long short-term memory (LSTM) networks.
6. Incorporating Satellite and IoT Data:

- Integrate satellite imagery data and IoT sensor data, such as soil moisture sensors and weather stations, to capture real-time information about crop health and environmental conditions.
- Satellite data can provide insights into vegetation indices and crop growth, while IoT data can offer detailed information on local environmental variables.

7. Validation and Model Evaluation:
   - Validate the model's performance on the validation set and use appropriate evaluation metrics, such as mean squared error (MSE) or root mean squared error (RMSE); after that, fine-tune the model if needed and perform cross-validation to ensure its generalizability.

8. Real-Time Predictions:
   - Once the model is trained and validated, it can be used for real-time crop yield predictions (by using algorithms like supervised learning algorithms) based on current weather conditions and other input variables.

9. Continuous Monitoring and Updating:
   - Continuously monitor the performance of the crop yield prediction model and update it regularly with new data.

7.2.2 Disease Detection and Pest Management

Utilizing computer vision and image analysis to detect diseases in crops has become a game-changer in modern agriculture. By harnessing the power of ML and image processing techniques, farmers can quickly and accurately identify diseases and pests that affect their crops, enabling them to take timely and targeted actions for disease management. Here's a brief explanation of how computer vision and image analysis are used for disease detection in crops:

1. Image Acquisition and Preprocessing:
   - High-resolution images of crops are captured using various sources, such as drones, satellites, or mounted cameras.
   - Preprocessing techniques are applied to enhance image quality, remove noise, and correct any image distortions or artifacts.

2. Segmentation of Plant Regions:
   - Computer vision algorithms are employed to segment the plant regions of interest from the background in the images.

- This step ensures that only the relevant parts of the image are analyzed, focusing on the leaves or affected areas.

3. Feature Extraction:
   - Relevant features, such as color, texture, shape, and size of the plant regions, are extracted from the segmented images.
   - These features serve as essential inputs for disease classification models.

4. Disease Classification:
   - ML algorithms, such as SVM, convolutional neural networks (CNN), or decision trees, are trained using labeled datasets of healthy and diseased crops.
   - The models learn to differentiate between healthy and infected crops based on the extracted features.

5. Disease Identification and Diagnosis:
   - Once the model is trained, it is deployed to predict disease presence in new, unseen images of crops.
   - The model classifies the crop as healthy or identifies the specific disease or pest affecting it.

6. Early Detection and Monitoring:
   - Continuous monitoring of crops using computer vision enables early detection of diseases before they spread extensively.

7.3 SOIL HEALTH AND NUTRIENT MANAGEMENT

7.3.1 Soil Fertility Analysis

Analyzing soil fertility using ML involves a series of steps to process and analyze soil data, build predictive models, and interpret the results. Here's a step-by-step process for soil fertility analysis using ML:

1. Data Collection:
   - Gathering soil samples from various locations in the field, ensuring they are representative of the entire area, which gives relevant information about each sample, such as soil type, pH, organic matter content, nutrient levels (nitrogen, phosphorus, potassium, etc.), and other soil properties.

2. Data Preprocessing:

- Cleaning the data by handling missing values and removing any outliers.
- Normalizing the numeric features to ensure that they have the same impact on the model.

3. Feature Selection:
   - Using techniques such as correlation analysis or feature importance from ML models to select relevant features.

4. Building the ML Model:
   - Choosing a suitable ML algorithm for soil fertility prediction. Common algorithms include linear regression, decision trees, random forests, SVM, and gradient boosting machines (GBM). We can split the data into training and testing sets to evaluate the model's performance.

5. Model Training:
   - Adjust hyperparameters to optimize the model's performance. We can use techniques like grid search or random search for hyperparameter tuning.

6. Model Evaluation:
   - Evaluate the model's performance on the testing set using appropriate evaluation metrics such as Mean Absolute Error (MAE), RMSE, or R-squared.
   - Compare the results with domain-specific thresholds to determine the model's accuracy.

7. Interpretation of Results:
   - Analyze the model's coefficients or feature importance to understand which soil properties have the most significant influence on soil fertility. And interpret the predictions to identify areas with low or high soil fertility.

8. Validation and Refinement:
   - Validate the model's performance using new soil samples collected independently from the training and testing sets.

9. Deploying the Model:
   - Integrate the model into an application or system that allows users to input soil data and receive fertility predictions.

10. Continuous Monitoring and Updates:
    - Continuously monitor the model's performance and update it regularly with new data to improve accuracy and adapt to changing soil conditions.

7.3.2 Irrigation Optimization

The implementation of sensor networks to monitor soil moisture levels and weather conditions is represented in Figure 7.2.

Employing ML algorithms to create intelligent irrigation systems, as shown in Figure 7.3.

Reducing water waste and ensuring adequate hydration for crops, as mentioned in Figure 7.4.

The role of reinforcement learning in dynamically adjusting irrigation schedules is shown in Figure 7.5.

7.4 LIVESTOCK MANAGEMENT

7.4.1 Automated Monitoring of Livestock

Automated monitoring of livestock is a technology-driven approach that involves using various sensors, devices, and data analytics to continuously monitor and track the health, behavior, and well-being of livestock in agricultural settings. This advanced monitoring system can provide real-time information to farmers, allowing them to identify and address potential issues promptly. Here's how automated monitoring of livestock works (Figure 7.6):

1. Wearable Devices and Sensors:
   - Livestock wear devices such as GPS-enabled collars, ear tags, or leg bands containing sensors.
   - These sensors can monitor vital signs like body temperature, heart rate, respiration rate, and activity levels of the animals.
2. Tracking and Localization:
   - GPS-enabled devices track the location and movements of livestock in the field or grazing areas.
   - This helps farmers keep track of their animals' whereabouts, prevent straying, and ensure efficient grazing management.
3. Health and Behavior Monitoring:
   - Sensors continuously monitor the health and behavior of the animals.
   - Anomalous behavior, such as decreased activity or changes in feeding patterns, can indicate health issues or stress.

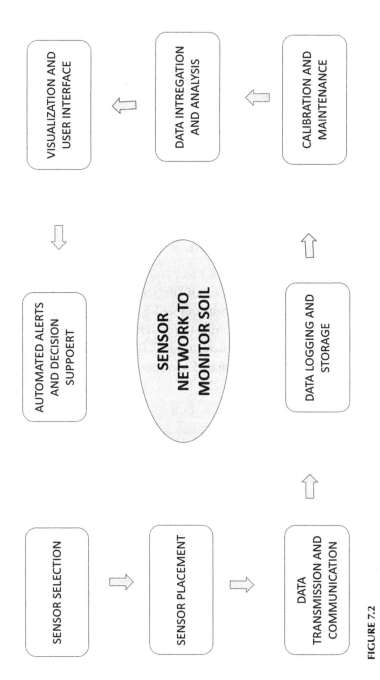

FIGURE 7.2

Implementation of sensor networks.

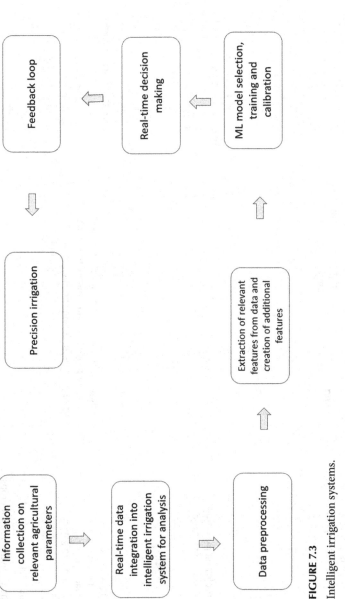

FIGURE 7.3

Intelligent irrigation systems.

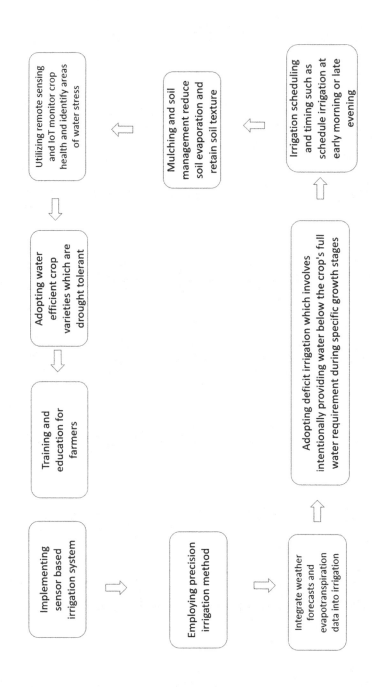

FIGURE 7.4

Adequate hydration for crops.

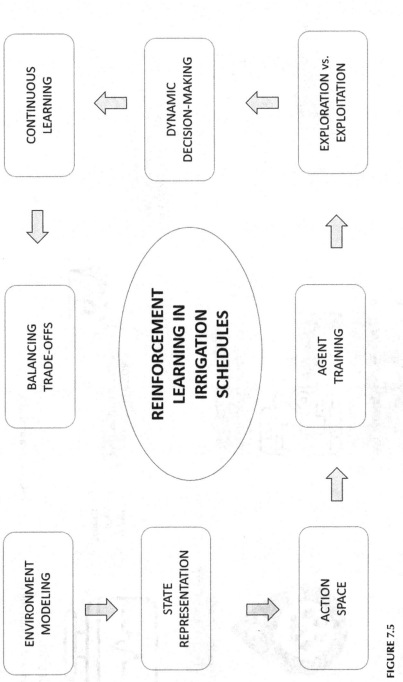

FIGURE 7.5

Dynamical adjustment of irrigation schedules.

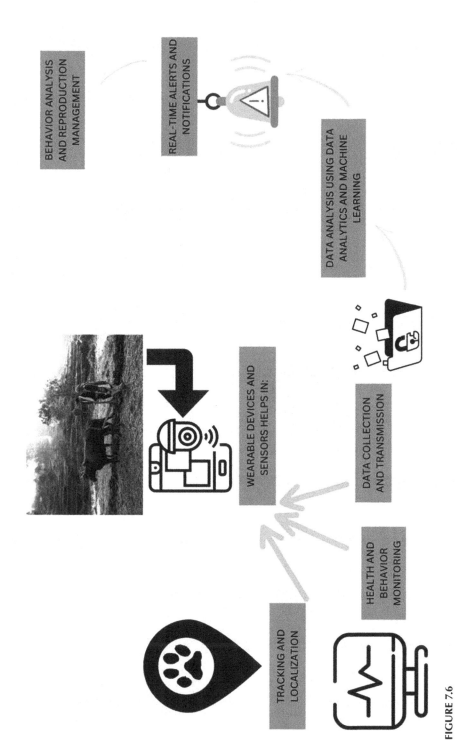

FIGURE 7.6

Automated monitoring using sensors, devices, and data analytics.

4. Data Collection and Transmission:
   - The sensor devices collect data from the livestock, and the information is transmitted wirelessly to a central data collection system.
   - Data may be sent to cloud-based platforms or local servers for storage and analysis.
5. Data Analytics and ML:
   - Data from multiple livestock are analyzed using data analytics and ML algorithms.
   - These algorithms can identify patterns and trends that may indicate specific health conditions or stress factors.
6. Real-Time Alerts and Notifications:
   - When abnormal data or predefined thresholds are detected, the system generates real-time alerts or notifications.
   - Farmers receive these alerts on their smartphones or computers, allowing them to take immediate action.
7. Remote Monitoring and Management:
   - Farmers can access the monitoring system remotely to check the health status and location of their livestock.
   - This feature is particularly useful for large-scale livestock operations or when animals are grazing in distant areas.
8. Behavior Analysis and Reproduction Management:
   - Automated monitoring can help in detecting estrus behavior in female animals, assisting in proper timing for breeding.
   - Behavior analysis may also identify signs of distress or disease early on, enabling timely intervention.

7.5 FEED OPTIMIZATION

Feed optimization using ML is an innovative approach in agriculture to efficiently manage animal nutrition and reduce feed costs while maximizing animal health and productivity. ML algorithms can analyze large amounts of data to create personalized and optimized feeding plans for livestock. Here's how feed optimization using ML works:

1. Data Collection:
   - Gather data on animal characteristics, such as species, breed, age, weight, and physiological status (e.g., lactating, gestating).

- Collect nutritional information about various feed ingredients, including nutrient composition and digestibility.

2. Nutrient Requirements:
   - Define the nutritional requirements for different classes of animals, considering their growth stages, reproductive status, and production goals.
   - Nutrient requirements are typically based on scientific guidelines and established feeding standards.

3. Formulating the Problem:
   - Convert the feed optimization problem into an ML problem by formulating it as an optimization task.
   - Define the objective function, which could be maximizing animal productivity (e.g., milk yield, weight gain) while minimizing feed costs.

4. Data Preprocessing:
   - Clean and preprocess the data, handling missing values and outliers to ensure data quality.

5. Feature Engineering:
   - Extract relevant features from the data, such as animal characteristics and nutrient compositions of feed ingredients.
   - Create composite features or ratios that represent the nutritional balance required for specific animal classes.

6. Model Selection:
   - Choosing appropriate ML algorithms based on specific feed optimization problems like regression, linear programming, and genetic algorithms, which are commonly used for feed optimization tasks.

7. Model Training:
   - Train the ML model on historical feeding data, which includes feed formulations and corresponding animal performance data, which learns to identify optimal feed combinations that meet the animals' nutritional requirements.

8. Model Validation and Evaluation:
   - Validating the ML model's performance on a separate dataset or through cross-validation we can evaluate the models ability to generate cost-effective feed formulations while meeting animal nutritional needs.

9. Real-Time Feed Formulation:

- Deploying the trained ML model in an automated feed formulation system, which is based on real-time data about animals and available feed ingredients and generates personalized and optimized feed plans.

10. Continuous Learning and Updates:

- Continuously monitoring the animal performance and feed costs to gather new data, which helps in regular updating of the ML model to improve its accuracy and adapt to changing feed ingredient availability or nutritional requirements.

7.6 CONCLUSION

ML is transforming agriculture by providing data-driven insights and decision-making capabilities. Precision farming, crop management, soil health analysis, and livestock management are just a few areas where ML applications are making a significant impact. As technology continues to advance, the agriculture industry stands to benefit even more from these innovations, leading to sustainable farming practices, increased yields, and improved food security. However, challenges such as data privacy, adoption barriers, and the digital divide must be addressed to ensure the widespread adoption of these transformative technologies. By combining domain expertise with ML capabilities, agriculture is poised to embrace a new era of productivity and sustainability.

BIBLIOGRAPHY

Ahmad, Latief, and Firasath Nabi. *Agriculture 5.0: Artificial Intelligence, IoT and Machine Learning.* CRC Press, 2021.

Alreshidi, E. Smart sustainable agriculture (SSA) solution underpinned by internet of things (IoT) and artificial intelligence (AI). (2019). arXiv preprint arXiv:1906.03106.

Anitha Mary, X., Popov, V., Raimond, K., Johnson, I., and Vijay, S. J. Scope and recent trends of artificial intelligence in Indian agriculture. *The digital agricultural revolution: Innovations and challenges in agriculture through technology disruptions.* 1–24, Wiley, 2022.

Dakir, A., F. Barramou, and O. B. Alami. Opportunities for artificial intelligence in precision agriculture using satellite remote sensing. *Geospatial Intelligence: Applications and Future Trends.* 107–117, 2022. https://doi.org/10.1007/978-3-030-80458-9_8

Javaid, M.A. Haleem, I. H. Khan, and R. Suman. Understanding the potential applications of artificial intelligence in agriculture sector. *Advanced Agrochem* 2.1. (2023): 15–30.

Jung, J., M. Maeda, A. Chang, M. Bhandari, A. Ashapure, and J. Landivar-Bowles. The potential of remote sensing and artificial intelligence as tools to improve the resilience of agriculture production systems. *Current Opinion in Biotechnology* 70 (2021): 15–22.

Khan, Yusuf Mustafa, and Sandeep Kumar. Applications of artificial intelligence in agriculture. *International Research Journal of Modernization in Engineering Technology and Science* 3 (2021): 1398–1402.

Khandelwal, Paras M., and Himanshu Chavhan. *Artificial intelligence in agriculture: An emerging era of research*. ResearchGate Publication, 2019.

Linaza, M. T., J Posada, J. Bund, P. Eisert, M. Quartulli, J. Döllner, A. Pagani, I. G. Olaizola, A. Barriguinha, T. Moysiadis, and L. Lucat. Data-driven artificial intelligence applications for sustainable precision agriculture. *Agronomy* 11 (2021): 1227.

Megeto, G. A. S., A. G. D. Silva,R. F. Bulgarelli, C. F. Bublitz, A. C. Valente, and D. A. G. D. Costa. Artificial intelligence applications in the agriculture 4.0. *Revista Ciência Agronômica* 51 (2021): e20207701.

Mekonnen, Y., S. Namuduri, L. Burton, A. Sarwat, and S. Bhansali. Machine learning techniques in wireless sensor network based precision agriculture. *Journal of the Electrochemical Society* 167.3 (2019): 037522.

Misra, N. N., Yash Dixit, Ahmad Al-Mallahi, Manreet Singh Bhullar, Rohit Upadhyay, and Alex Martynenko. IoT, big data, and artificial intelligence in agriculture and food industry. *IEEE Internet of Things Journal* 9 (2020): 6305–6324.

Mitra, Alakananda, Sukrutha L. T. Vangipuram, Anand K. Bapatla, Venkata K. V. V. Bathalapalli, Saraju P. Mohanty, Elias Kougianos, and Chittaranjan Ray. "Everything you wanted to know about smart agriculture." (2022). arXiv preprint arXiv:2201.04754.

Panpatte, D. G. *Artificial intelligence in agriculture: An emerging era of research*. Anand Agricultural University, 2018.

Sarkar, M. R.S. R. Masud, M. I. Hossen, and M. Goh. A comprehensive study on the emerging effect of artificial intelligence in agriculture automation. In *2022 IEEE 18th International Colloquium on Signal Processing & Applications (CSPA)*, 2022 May 12, pp. 419–424). IEEE.

Shadrin, D., A. Menshchikov, D. Ermilov, and A. Somov. Designing future precision agriculture: Detection of seeds germination using artificial intelligence on a low-power embedded system. *IEEE Sensors Journal* 19.23. (2019): 11573–11582.

Sharma, A., M. Georgi, M. Tregubenko, A. Tselykh, and A. Tselykh Enabling smart agriculture by implementing artificial intelligence and embedded sensing. *Computers & Industrial Engineering* 165 (2022): 107936.

Singh, Garima, Anamika Singh, and Gurjit Kaur. Role of artificial intelligence and the internet of things in agriculture. In *Artificial Intelligence to Solve Pervasive Internet of Things Issues*, pp. 317–330. Academic Press, 2021a.

Singh, P., and A. Kaur. A systematic review of artificial intelligence in agriculture. *Deep Learning for Sustainable Agriculture*. 57–80, 2022. https://doi.org/10.1016/B978-0-323-85214-2.00011-2

Singh, Rajesh, and Anita Gehlot, Mahesh Kumar Prajapat, and Bhupendra Singh. *Artificial Intelligence in Agriculture*. CRC Press, 2021b.

Singh, Ritesh Kumar, Rafael Berkvens, and Maarten Weyn. AgriFusion: An architecture for IoT and emerging technologies based on a precision agriculture survey. *IEEE Access* 9 (2021c): 136253–136283.

Siregar, R. R. A., K. B. Seminar, S. Wahjuni, and E. Santosa. Vertical farming perspectives in support of precision agriculture using artificial intelligence: A review. *Computers* 11.9. (2022): 135.

Sishodia, R. P. R. L. Ray, and S. K. Singh. Applications of remote sensing in precision agriculture: A review. *Remote Sensing* 12 (2020):3136.

Skvortsov, E. A. Prospects of applying artificial intelligence technologies in the regional agriculture. *Ekonomika Regiona= Economy of Regions* 2 (2020): 563.

Sood, Amit, Rajendra Kumar Sharma, and Amit Kumar Bhardwaj. Artificial intelligence research in agriculture: A review. *Online Information Review* 46.6 (2022): 1054–1075.

Subeesh, A., and C. R. Mehta. Automation and digitization of agriculture using artificial intelligence and internet of things. *Artificial Intelligence in Agriculture* 5 (2021): 278–291.

Tzachor, Asaf, Medha Devare, Brian King, Shahar Avin, and Seán Ó hÉigeartaigh. Responsible artificial intelligence in agriculture requires systemic understanding of risks and externalities. *Nature Machine Intelligence* 4.2 (2022): 104–109.

Yousaf, A., V. Kayvanfar, A. Mazzoni, and A. Elomri. Artificial intelligence-based decision support systems in smart agriculture: Bibliometric analysis for operational insights and future directions. *Frontiers in Sustainable Food Systems.* 6 (2023): 1053921.

8

Using Deep Learning to Detect Apple Leaf Disease

Syed Nisar Hussain Bukhari, Rukaya Manzoor, Ummer Iqbal, and Muneer Ahmad Dar
National Institute of Electronics and Information Technology (NIELIT), Srinagar, India

8.1 INTRODUCTION

Kashmir, land rich in its culture and languages, is also rich in its vegetation. Its soil supports the growth of a variety of fruits, the apple being the most famous. Apple production brings in a lot of wealth. The apple tree originated in central Asia and is now cultivated worldwide. Its botanical name is *malus domestica* [1]. In Kashmir, a wide variety of apples are cultivated viz Delicious, Golden, Ambri, Chamboora, etc. Apples are grown in almost every district of Kashmir and generate annual revenue of Rs. 1500 crores [2]. The apple tree is, however, vulnerable to certain kinds of disease, which affect its leaves. The typical apple leaf diseases found in Kashmir are Alternaria leaf blotch, apple scab, and apple rot, as shown in Figure 8.1. These diseases render the whole orchard devastated, which is a big challenge for the farmers. These diseases damage the quality and decrease the quantity of crops. The traditional ways to diagnose the disease cause greater damage, as the methods are slow, labor-intensive, and nonreliable. These methods need experts with extensive knowledge and experience and are often inaccurate. The pesticide that is then used is also improper and becomes an economic and environmental burden.

Now it has become the need of the hour to change the traditional time-consuming methods to novel technology-based methods. The advanced technology has put machine learning in use, which automatically detects and classifies the disease of apple leaves. However, this method has a

DOI: 10.1201/9781003485179-8

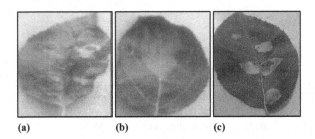

(a)   (b)   (c)

FIGURE 8.1

Typical apple diseases: (a) Alternaria leaf blotch, (b) apple scab, and (c) apple rot.

drawback, as it cannot be applied to raw data and requires preprocessing. Deep learning has now emerged as a better alternative, as it can be applied to raw data, and no extra step of preprocessing is needed. Further advancements in deep learning have led to the idea of a convolution neural network (CNN). It is particularly used for image classification. CNN comprises two components – viz, convolution and pooling used for feature extraction, and the classifier component, which predicts the output. In this study, various versions of CNN algorithms have been used for predicting and classifying different types of disease in apple leaves.

This chapter contains four sections. The first section is about related work done in related fields. The second section talks about experimental details. This section includes methods, model building, data set, and experimental environment. The third section is about analyzing the results using different performance evaluation metrics. The last section sums up the conclusion of the study.

8.2 RELATED WORK

Deep learning and machine learning techniques have been practiced on different disease predictions on different types of agricultural products. The research related to the identification and classification of plant leaf diseases is presented in this section. The remainder of this section gives a detailed explanation of the research undertaken in this field. The summarization of this section is given in Table 8.1.

Zhang et al. [3] introduced a new method of apple leaf disease identification. The model was based on the support vector machine (SVM) classifier. The dataset for the identification of diseased apple leaves contains 90

TABLE 8.1

Related Work Done in Related Fields

Author	Year	Description	Algorithm	Classification Accuracy (%)
Zhang et al. [3]	2017	Identification of apple leaf diseases	SVM	90
Dubey et al. [4]	2012	Identification and categorization of apple fruit disease	SVM	93
Arivazhagan et al. [5]	2013	Detection and classification of leaf disease	SVM	94
Brahimi et al. [6]	2017	Recognition of tomato leaf diseases	CNN	99
Khan et al. [7]	2021	Classification and identification of apple diseases	CNN	97.18
Jan et al. [8]	2020	Apple disease diagnosis for predicting apple scab and leaf/spot blight disease	MLP	99.1

images of healthy apple leaves, mosaic leaf, powdery mildew leaves, and rust leaves. This approach acquired an identification accuracy of more than 90%. S. R. Dubey et al. [4] developed a system for the detection and classification of apple fruit disease. In this system, image segmentation was done using K means clustering. The segmented images were then used to obtain the art features. Finally, the images are classified into apple blotch, apple rot, and apple scab using multiclass SVM. This model achieved an accuracy of 93%. Arivazhagan et al. [5] developed a software-based automatic recognition and classification of plant leaf disease. For the processing part, color conversion and segmentation were applied to images. The system used an SVM classifier for recognition and classification. An accuracy of 94% was achieved. Brahimi et al. [6] also applied a deep-learning (CNN algorithm) model for a similar study classifying tomato disease. The model used leaf images and achieved a classification accuracy of 99%. Khan et al. [7] used the concept of CNN to perform feature extraction from leaf images. The CNN model worked to predict the diseased leaves and acquired an accuracy of about 97%. Jan et al. [8] developed an outstanding apple disease recognition system to recognize leaf/spot blight and apple scab disease. The proposed system used shape-based and low-level features of leaves to train a multilayer perception (MLP) model. The model achieves an accuracy of 99.1%.

8.3 PROPOSED METHODOLOGY

The methodology employed in this study is a structured approach designed to comprehensively address the challenges of predicting and classifying apple leaf diseases. These steps encompass acquiring an image dataset, performing image preprocessing, constructing a predictive model, evaluating the model's performance, and analyzing the results. The methodology is shown in the following flow chart (Figure 8.2).

8.3.1 Methods

The study has been conducted using a custom CNN, VGG 16, ResNet-50 and Inception V3. The pretrained networks VGG 16, ResNet-50, and Inception V3 are all powered by convolution neural techniques. The working of CNN, along with its mathematical modeling, is explained next.

CNN is a type of deep-learning algorithm that is mostly used for image classification [9]. The architecture of CNN contains many layers, and each layer performs a different function. The layers of CNN are connected to each other, and the output of a layer is input for the next layer [10]. Mainly, there are three layers in CNN – namely, the convolution layer, pooling layer, and fully connected layer.

8.3.1.1 Convolution Layer

The convolution layer is a key layer in CNN, whose function is feature extraction [11]. Feature extraction is done by using a filter or kernel. In the convolution layer, an activation/feature map is formed by the convolution operation between the input image and the kernel.

Equation 8.1 represents the convolution operation mathematically.

$$O_j = f\left(\sum_{i=1}^{n} a_i * k_{i,j} + b_j\right) \tag{8.1}$$

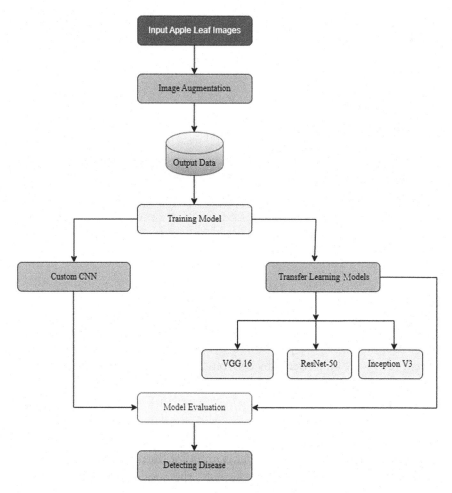

FIGURE 8.2

Methodological steps.

First of all, the input image a_i is convoluted with the kernel $k_{i,j}$; then the summation of convoluted operation is calculated, and a bias b_j is added with the resultant. Lastly, an activation function f is applied to the resultant, and the feature map O_j is produced.

Bias is a constant value that is used within an activation function in order to shift the position of the curve left or right to delay or accelerate the firing of a neuron [12].

The function of the activation function is to decide whether the neuron is fired or not. The most widely used activation is the Rectified Linear Unit (ReLU). The advantage of the ReLU function is that it does not activate all the neurons simultaneously. The ReLU function does not fire the neuron if the input values are negative [13].

Equation 8.12 represents the ReLU function mathematically.

$$g(x) = \max(0, x) \tag{8.2}$$

8.3.1.2 Pooling Layer

The pooling layer is present after the convolution layer in CNN. The function of the pooling layer is to reduce the dimensions of the feature map. In pooling, the neighboring elements of convolution output are combined [11]. Average pooling and max pooling are the main pooling operations.

Average pooling is a pooling operation that filters the feature map and selects the average value from the filtered patch [14], mathematically represented in Equation 8.3.

$$f_{avg}(x) = \frac{1}{n} \sum_{i=1}^{n} x_i \tag{8.3}$$

Max pooling is a pooling operation that filters the feature map and selects the maximum value from the filtered patch [14]. Equation 8.4 gives the mathematical representation of this.

$$f_{max}(x) = \max(x_i) \tag{8.4}$$

8.3.1.3 Fully Connected Layer

After a number of convolution and pooling layers, there are a number of fully connected layers. The output (feature map) from the convolution and pooling layers are joined with each other to form a larger feature map. In another way, a fully connected layer flattens the input image so that it converts the multidimensional input image into a one-dimensional array (vector) [10]. Each fully connected layer has an activation function. Mostly for multiclass image classification, the Softmax activation function is used for the last fully connected layer. Softmax function converts the one-dimensional array into the probability distribution that sums up to 1 [15].

Equation 8.5 represents the Softmax activation function mathematically.

$$S = \frac{\exp^{zi}}{\sum_{j=1}^{k} \exp^{zj}} \qquad (8.5)$$

8.3.2 Model Building

The CNN algorithms used in this study for model building are given in the next section:

8.3.2.1 Custom CNN

The proposed custom CNN is an 11-layer CNN containing 4 convolution layers, 4 max-pool layers, and 3 dense layers. The input size of 150 × 150 is passed to the 11 layers of the network. This model uses the ReLU activation function for hidden layers and the Softmax activation function for the output layer. Figure 8.3 shows the architecture of the model.

8.3.2.2 VGG16

VGG (visual geometric group), proposed by Zisserman et al. in the year 2014, is a 16-layer CNN working on 13 convolution layers, 5 max-pooling layers, and 3 dense layers that add up to 21 layers, out of which only 16 are weighted [16]. The architecture is shown in Figure 8.4. The input size of 224 × 224 is passed to 16 layers of the network [17].

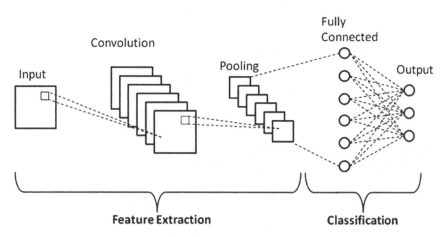

FIGURE 8.3
Architecture of custom CNN.

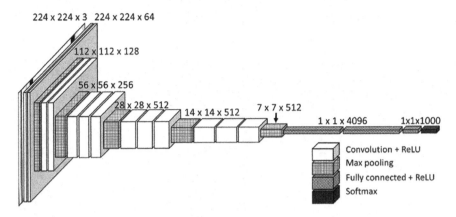

FIGURE 8.4
Architecture of VGG16.

8.3.2.3 ResNet-50

ResNet (residual network), proposed by Kaiming et al. in the year 2015, is a 50-layer CNN with 48 convolution layers, 1 max-pool layer, and 1 average pool layer [18]. The architecture of Resnet-50 is shown in Figure 8.5. The input size of 180 × 180 is passed to the 50 layers of the network [19].

8.3.2.4 InceptionV3

Inception v3 is proposed by Szegedy et al. in the year 2015 [20]. The inception V3 contains 5 convolution layers, 2 max-pooling layers, 11 inception modules, 1 average pooling layer, and 1 dense layer [21]. The architecture of Inception V3 is shown in Figure 8.6. The input size of 229 × 229 is passed to the 48 layers of the network.

8.3.3 Dataset

The dataset for this study was collected from the dataset repository Kaggle under the title "Kashmir Apple Leaf Disease" [22]. The data set contains 419 images of infected and healthy leaves. Table 8.2 contains the number of images for individual classes.

8.3.4 Data Augmentation

Data augmentation is a technique by which artificial data is generated from the dataset in order to decrease overfitting during the training of a model. In augmentation, geometric and color space transformations

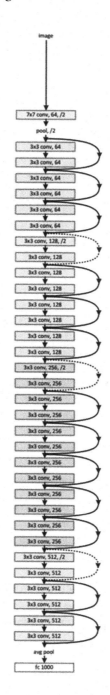

FIGURE 8.5

Architecture ResNet-50.

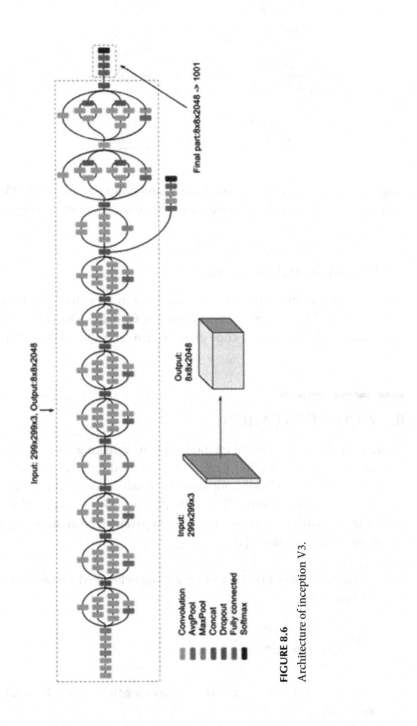

FIGURE 8.6

Architecture of inception V3.

TABLE 8.2

Number of Records in Dataset

Class	Total Number of Images
Healthy	46
Apple Rot	103
Alternaria leaf Blotch	111
Apple Scab	159

(flipping, resizing, cropping, brightness, and contrast) are used [23]. In this study, 419 image files of the dataset have been augmented to generate 2,000 files.

8.3.5 Experimental Environment

Coding for experimentation is performed within a Python environment called Google Colab – a product developed by Google Research. It enables the writing and execution of code directly through a web browser [24].

8.4 MODEL EVALUATION

Model evaluation refers to the methodology of evaluating the model's performance. Several evaluation parameters are used for that purpose that are all based on the following outcomes of classification predictions: True Negative (TN), True Positive (TP), False Positive (FP), and False Negative (FN). Conventional performance evaluation parameters are accuracy, precision, recall, and F1-score [25].

- Accuracy: The proportion of predictions classified correctly by the model.

$$Acc = \frac{TN + TP}{TN + TP + FP + FN} \tag{8.6}$$

- Precision: The ratio of positive values classified correctly to all the positives.

TABLE 8.3

Evaluation Parameters

Algorithm	Classification Accuracy (%)	Precision	Recall	F1-Score
Custom CNN	80	0.78	0.77	0.77
VGG16	85	0.82	0.81	0.81
ResNet-50	81	0.79	0.77	0.78
InceptionV3	0.83	0.80	0.78	0.79

$$\text{Precision} = \frac{\text{TP}}{\text{TP} + \text{FP}} \tag{8.7}$$

- Recall: The ratio of correctly classified positive values to the total positives in a class.

$$\text{Recall} = \frac{\text{TP}}{\text{TP} + \text{FN}} \tag{8.8}$$

- F1-score: Harmonic mean (weighted average) of values given returned by recall and precision.

$$\text{F1} - \text{score} = \frac{2*\left(\text{Precision}*\text{Recall}\right)}{\text{Precision} + \text{Recall}} \tag{8.9}$$

The values of performance evaluation parameters for all the models are represented in Table 8.3.

8.5 RESULTS

In the current study, four different models – custom CNN, VGG16, ResNet-50, and InceptionV3 – were employed for predicting and classifying apple leaf diseases. The performance metrics, including classification accuracy, precision, recall, and F1-score, were used to evaluate the models [26–30]. Among these models, VGG16 emerged as the top performer, exhibiting an impressive accuracy of 85%. The precision, recall, and F1-score for VGG16 were also commendable, with values of 0.82, 0.81, and 0.81, respectively. The values of the performance parameters of all the models are shown in Table 8.3. This suggests that VGG16 excelled in correctly identifying and categorizing apple leaf diseases, showcasing its potential for practical implementation in disease detection systems for apple orchards.

Furthermore, the comparative analysis of the models highlight the significance of choosing an appropriate deep-learning architecture. While custom CNN, ResNet-50, and InceptionV3 demonstrated respectable performance, VGG16 consistently outperformed them across all evaluation metrics. The robustness of VGG16 in handling the complexities of apple leaf images and accurately predicting diseases underscores its suitability for real-world applications. These results not only validate the effectiveness of deep learning in disease prediction but also emphasize the importance of model selection, with VGG16 standing out as the most reliable and accurate choice for the detection of apple leaf diseases in the context of Kashmir's orchards.

8.6 CONCLUSION

At the beginning of this chapter, the advancements of machine learning and deep learning and the purpose of this study were discussed. The next segment was about the work done in this regard. Following that, the experiment setup was discussed, which comprised the dataset, steps followed, and models [31–35]. The last section is about the results of the models performed using various performance evaluation parameters.

This study presents a significant advancement in the field of agriculture, specifically addressing the challenges faced by apple cultivators in Kashmir. The proposed deep learning–based system, utilizing custom CNN, VGG16, ResNet-50, and InceptionV3 models, exhibits a commendable accuracy of 85%, with VGG16 leading in performance. The classification of apple leaves into distinct categories – healthy, apple rot, Alternaria leaf blotch, and apple scab – proves to be an invaluable contribution to the farming community. The study not only offers an efficient and cost-effective alternative to traditional disease prediction methods but also signifies a paradigm shift toward leveraging advanced technologies for agricultural sustainability.

Looking forward, the future of this study lies in continuous refinement and expansion. Optimization of existing models, exploration of ensemble techniques, and incorporation of real-time data sources could further enhance the accuracy and applicability of the predictive system. Extending the scope beyond apple leaves to encompass a broader range of crops would amplify the impact of these models in diverse agricultural contexts. Collaborative efforts involving experts in agriculture, data science, and technology innovation could drive interdisciplinary research, fostering innovative solutions for crop disease management. The integration of Internet of Things (IoT) devices, sensors, and big data analytics holds promise for proactive disease control strategies, ensuring the resilience and prosperity of farming communities. This study marks a pivotal step toward the integration of cutting-edge technologies in agriculture, with far-reaching implications for the future of sustainable and technology-driven farming practices.

REFERENCES

[1] A. Cornille, P. Gladieux, M. J. M. Smulders, I. Roldán-Ruiz, F. Laurens, B. le Cam, A. Nersesyan, J. Clavel, M. Olonova, L. Feugey, I. Gabrielyan, X. G. Zhang, M. I. Tenaillon, and T. Giraud. New Insight into the History of Domesticated Apple: Secondary Contribution of the European Wild Apple to the Genome of Cultivated Varieties. *PLoS Genetics*, 8(5): e1002703, 2012. doi: 10.1371/journal.pgen.1002703

[2] *Apple Industry in J&K A Tumbling Sector – The Kashmir Horizon*. (n.d.). Retrieved January 22, 2023, from https://thekashmirhorizon.com/2022/09/27/apple-industry-in-jk-a-tumbling-sector/

[3] Z. Chuanlei, Z. Shanwen, Y. Jucheng, S. Yancui, and C. Jia. Apple Leaf Disease Identification Using Genetic Algorithm and Correlation Based Feature Selection Method. *International Journal of Agricultural and Biological Engineering*, 10(2): 74–83, 2017.

[4] S. R. Dubey and A. S. Jalal. Detection and Classification of Apple Fruit Diseases Using Complete Local Binary Patterns. In *2012 Third International Conference on Computer and Communication Technology*, Allahabad, India, 2012, pp. 346–351. doi: 10.1109/ICCCT.2012.76

[5] S. Arivazhagan, R. N. Shebiah, S. Ananthi, and S. V. Varthini. Detection of Unhealthy Region of Plant Leaves and Classification of Plant Leaf Diseases Using Texture Features. *Agricultural Engineering International: CIGR Journal*, 15(1): 211–217, 2013.

[6] M. Brahimi, K. Boukhalfa, and A. Moussaoui. Deep Learning for Tomato Diseases: Classification and Symptoms Visualization. *Applied Artificial Intelligence*, 31(4): 299–315, 2017. doi: 10.1080/08839514.2017.1315516

[7] A. I. Khan, S. M. K. Quadri, and S. Banday. Deep Learning for Apple Diseases: Classification and Identification. *International Journal of Computational Intelligence Studies*, 10(1): 1–12, 2021.

[8] M. Jan and H. Ahmad. Image Features Based Intelligent Apple Disease Prediction System: Machine Learning Based Apple Disease Prediction System. *International Journal of Agricultural and Environmental Information Systems (IJAEIS)*, *11*(3): 31–47, 2020.

[9] V. A. Kherdekar Convolution Neural Network Model for Recognition of Speech for Words used in Mathematical Expression. *Turkish Journal of Computer And Mathematics Education*, *12*(6): 4034–4042, 2021.

[10] R. Prasath, M. Anand, and S. Hariharan. Image Classification Using Convolutional Neural Networks. (n.d.). Retrieved from http://www.acadpubl.eu/hub/

[11] Q. Li, W. Cai, X. Wang, Y. Zhou, D. D. Feng, and M. Chen. Medical Image Classification with Convolutional Neural Network. In *2014 13th International Conference on Control Automation Robotics and Vision, ICARCV 2014*, 2014, pp. 844–848. doi: 10.1109/ICARCV.2014.7064414

[12] *Importance of Neural Network Bias and How to Add It.* n.d. Retrieved March 3, 2023, from https://www.turing.com/kb/necessity-of-bias-in-neural-networks

[13] *ReLU Activation Function Explained | Built In.* n.d. Retrieved March 3, 2023, from https://builtin.com/machine-learning/relu-activation-function

[14] *Comprehensive Guide to Different Pooling Layers in Deep Learning.* n.d. Retrieved March 3, 2023, from https://analyticsindiamag.com/comprehensive-guide-to-different-pooling-layers-in-deep-learning/

[15] *Softmax Activation Function: Everything You Need to Know|Pinecone.* n.d. Retrieved March 3, 2023, from https://www.pinecone.io/learn/softmax-activation/

[16] *Understanding VGG16: Concepts, Architecture, and Performance.* n.d. Retrieved January 24, 2023, from https://datagen.tech/guides/computer-vision/vgg16/

[17] *The Architecture of the VGG-16 CNN Model [26].|Download Scientific Diagram.* n.d. Retrieved January 24, 2023, from https://www.researchgate.net/figure/The-architecture-of-the-VGG-16-CNN-model-26_fig5_326349019

[18] *ResNet-50: The Basics and a Quick Tutorial.* n.d. Retrieved January 24, 2023, from https://datagen.tech/guides/computer-vision/resnet-50/

[19] *ResNet50_From_Scratch_Tensorflow | This Repository Implements the Basic Building Blocks of Deep Residual Networks Which Is Trained on SIGNS Dataset to Detect Numbers On Hand] Images.* n.d. Retrieved January 24, 2023, from https://jananisbabu.github.io/ResNet50_From_Scratch_Tensorflow/

[20] *Inception-v3 Explained|Papers with Code.* n.d. Retrieved January 24, 2023, from https://paperswithcode.com/method/inception-v3

[21] C. Szegedy, V. Vanhoucke, S. Ioffe, and J. Shlens.*Rethinking the Inception Architecture for Computer Vision.* n.d.

[22] *Kashmiri Apple Plant Disease Dataset | Kaggle.* n.d. Retrieved January 23, 2023, from https://www.kaggle.com/datasets/hsmcaju/d-kap

[23] Johnson, M. n.d. *What Is Data Augmentation in a CNN? Python Examples.* Retrieved March 3, 2023, from https://nnart.org/what-is-data-augmentation-in-a-cnn/

[24] *Welcome to Colaboratory - Colaboratory.* n.d. Retrieved January 23, 2023, from https://colab.research.google.com/

[25] M. Hossin, and M. N. Sulaiman. A Review on Evaluation Metrics for Data Classification Evaluations. *International Journal of Data Mining & Knowledge Management Process*, *5*(2): 01–11, 2015. doi: 10.5121/ijdkp.2015.5201

[26] S. N. H. Bukhari, J. Webber, and A. Mehbodniya. Decision Tree Based Ensemble Machine Learning Model for the Prediction of Zika Virus T-Cell Epitopes as Potential Vaccine Candidates. *Scientific Reports, 12*: 7810, 2022. doi: 10.1038/s41598-022-11731-6

[27] G. S. Raghavendra, S. Shyni Carmel Mary, P. B. Acharjee, V. L. Varun, S. N. H. Bukhari, C. Dutta, and I. A. Samori. An Empirical Investigation in Analysing the Critical Factors of Artificial Intelligence in Influencing the Food Processing Industry: A Multivariate Analysis of Variance (MANOVA) Approach. *Journal of Food Quality, 2022*, Article ID 2197717, 7 pages, 2022. doi: 10.1155/2022/2197717

[28] S. N. H. Bukhari, A. Jain, E. Haq, A. Mehbodniya, and J. Webber. Ensemble Machine Learning Model to Predict SARS-CoV-2 T-Cell Epitopes as Potential Vaccine Targets. *Diagnostics, 11*(11): 1990, 2021. MDPI AG. doi: 10.3390/diagnostics11111990

[29] S. L. Bangare, D. Virmani, G. R. Karetla, P. Chaudhary, H. Kaur, S. N. H. Bukhari, and S. Miah. Forecasting the Applied Deep Learning Tools in Enhancing Food Quality for Heart Related Diseases Effectively: A Study Using Structural Equation Model Analysis. *Journal of Food Quality, 2022*, Article ID 6987569, 8 pages, 2022. doi: 10.1155/2022/6987569

[30] C. M. Anoruo, S. N. H. Bukhari, and O. K. Nwofor. Modeling and Spatial Characterization of Aerosols at Middle East AERONET Stations. *Theoretical and Applied Climatology, 152*: 617–625, 2023. doi: 10.1007/s00704-023-04384-6

[31] F. Masoodi, M. Quasim, S. Bukhari, S. Dixit, and S. Alam. *Applications of Machine Learning and Deep Learning on Biological Data.* CRC Press, 2023.

[32] S. N. H. Bukhari, A. Jain, and E. Haq. A Novel Ensemble Machine Learning Model for Prediction of Zika Virus T-Cell Epitopes. In Gupta, D., Polkowski, Z., Khanna, A., Bhattacharyya, S., and Castillo, O. (eds) *Proceedings of Data Analytics and Management. Lecture Notes on Data Engineering and Communications Technologies,* vol. 91. Springer, Singapore, 2022. doi: 10.1007/978-981-16-6285-0_23

[33] S. N. H. Bukhari, F. Masoodi, M. A. Dar, N. I. Wani, A. Sajad, and G. Hussain. Prediction of Erythemato-Squamous Diseases Using Machine Learning. In *Auerbach Publications eBooks*, pp. 87–96, 2023. doi: 10.1201/9781003328780-6

[34] S. N. H. Bukhari, A. Jain, E. Haq, A. Mehbodniya, and J. Webber. Machine Learning Techniques for the Prediction of B-Cell and T-Cell Epitopes as Potential Vaccine Targets with a Specific Focus on SARS-CoV-2 Pathogen: A Review. *Pathogens, 11*(2): 146, 2022. MDPI AG. doi: 10.3390/pathogens11020146

[35] S. Nisar, H. Bukhari, and M. A. Dar. Using Random Forest to Predict T-Cell Epitopes of Dengue Virus. *Adv. Appl. Math. Sci.. 20*(11): 2543–2547, 2021.

9

Agricultural Crop-Yield Prediction: Comparative Analysis Using Machine Learning Models

Kukatlapalli Pradeep Kumar, S. Babu Kumar,
Amarthya Dutta Gupta, Kevin Johnson,
and Meghan Mary Michael
CHRIST (Deemed to be University), Kengeri Campus, Bangalore, India

9.1 INTRODUCTION

The economy of emerging countries like India greatly benefits from agriculture, which is one of the most popular habitations. About 200.2 million hectares of land are used for agriculture in India, which is a significant source of income [1]. In India, 70% of the population relies on it for a living. Our economy is under pressure to increase food output because of overpopulation. In recent years, there has been a broad desire for growth across many industries due to the quick development of information technology [2]. One of the sectors with a lot of space for development is agriculture. This disadvantage is mostly due to the lack of a reliable agricultural model and enough farmer guidance. Implementing smart agricultural applications in isolated Indian villages is extremely difficult, especially when it comes to meeting community demands. For instance, illiterate and impoverished people cultivate rice using traditional ways.

Farmers will find it easier to make decisions if there is a mechanism that can automate irrigation, determine the quality of the soil, and forecast weather conditions, as well as price. Agriculture must be greatly advanced through innovations in this age of digitalization if it is to reach out to marginalized and small-scale farmers. Several harvests were lost each year as

DOI: 10.1201/9781003485179-9

a result of inadequate technical expertise and unanticipated patterns in the weather, such as temperature swings and rainfall, which had a considerable effect on agricultural yields and revenue.

Crop yield prediction is a crucial responsibility for agricultural factors. The management of maintaining the agricultural factors and ensuring the highest output depends heavily on crop yield prediction (CYP). The two most crucial feature sets are what CYP mostly focuses on. According to the farmer, one set of data includes the methods used for irrigation, fertilizer application, and field preparation. Environmental variables that are governed by nature, such as solar radiation, temperature, and rainfall, are included in a further set of data. Communication with field management methods and environmental factors affect how genetic markers behave. The sheer number of variables that must be taken into consideration makes it nearly difficult to predict crop yields with any degree of precision. Precision agriculture, which incorporates agricultural yield prediction, has several advantages, including enhancing crop output and quality while lowering environmental impact. Understanding the combined impacts of deficits in nutrients and water, illnesses, pests, and other variables in the field over the growing season is made easier by using crop yield simulations. This may result in environmentally conscious agriculture methods that also benefit society at large [3].

Crop yields may be harder to forecast due to interactions between genotypes and environmental variables. It is hard to accurately predict the complex nonlinear effect of outside influences, including the weather. Understanding the cumulative impacts of water and nutrient deficits, pests, illnesses, the influence of crop yield variability, and other variables in the field across the cultivation season is made easier with the assistance of crop yield simulations. Additionally, while making financial decisions, farmers and cultivators take crop projections into account. The two biggest obstacles in the current era of digital agriculture are collecting data and data integration.

Machine learning (ML), which is a subset of AI, concentrates on learning, and it is a spontaneous process that provides the best CYP, depending on a number of factors [4]. The ML approach analyzes a dataset to find information, patterns, and connections. The dataset should be used to train the algorithm so that the results may be predicted based on prior knowledge. The prediction technique is constructed using several characteristics; for instance, model parameters can be corroborated using data from the learning. Based on the presented features, ML is a practical technique that improves agricultural output forecasts. Numerous studies have used ML techniques for categorizing CYP, including random forests, regression trees, and multivariate regression. By analyzing past data, we may choose the prediction model's parameters during the later phase of training. The evaluation takes a fraction of the preceding data into account. Numerous techniques, including classification and clustering, are used for the prediction. Information is taken out of the data [5].

Agro-businesses can manage supply chain choices like production scheduling by using exact risk and crop production forecasts. Crop output forecasts serve as the cornerstone for marketing. Crop identification and mapping gain from the use of multitemporal imaging, which makes it possible to classify costs by accounting for variations in crop type reflectance. Using agricultural yield indicators to identify problems early and apply solutions might increase crop output and profitability. One new development is the use of hybrid models, which mix several ML methods to increase accuracy.

Artificial neural networks (ANNs) with the study of representations are used in deep learning, a branch of ML. In the presence of complex data, such as maturity groups and zones, genetic details, and various meteorological conditions, deep-learning methods may provide answers. The nonlinear relationships between projected yield alongside the multivariate data inputted (cluster information, meteorological factors, and maturity group) may be learned extremely effectively using this technique. The long-term temporary dependency in challenging multiple-variate sequences can be captured by the Long-Short Term Memory (LSTM) network, which is very beneficial for time series modeling. The prediction potential of ML models may also be increased by combining data from other sources, such as 3D mapping, sensor-based social condition data, and drone-based soil color data.

9.2 LITERATURE REVIEW

Crop yield is a topic that is of much economic and global importance. It can help manage agricultural resources efficiently and ensure food security. Advancements in CYP can boost the economy of a country and help in the decision-making process for farmers with respect to planting, harvesting, and resource allocation. These literary articles serve as the foundation for this chapter, which aims to come up with a model that can take the outcomes and observations from these studies further. This section of the chapter presents the preexisting studies conducted on CYP using various models such as deep learning, neural networks (NN), and ML models.

9.2.1 Neural Networks

Inam Ali et al. [6] find correlations between attributes from existing rice data and implement ML and analytics-based approaches to CYP. The study focuses on the data obtained from the district of Larkana in Pakistan, which hadn't previously been researched from the agricultural yield aspect. The agricultural sector in Pakistan contributes significantly to the economy of the country. The study found that by using the Apriori algorithm on the rice crop dataset of Larkana, researchers were able to share vital yield predictions, which could be broadly extended to other crops by altering the specific attributes. The data utilized for the study was collected from various local associated departments. The final rice yield dataset contained the targeted parameters: year, yield, humidity, wind speed, rainfall temperature, and area. The data collected contained the data for the aforementioned parameters for a duration of six years. The NN approach was used for the optimization of rice yield data, as it is very important and is preferred due to its level of accuracy being acceptable. NNs were used to optimize the results of rice yield prediction, which provides high-value yield that would be beneficiary to farmers, local agriculturalists, and government officials to make decisions based on rice crops. They can evaluate forecasting values of the connected variables outlined in the research and utilize this data, together with historical patterns, to make better data-driven decisions. The results of the study can strengthen the economy of the country by increasing rice production and also aid in preventing monetary losses incurred by farmers.

In the study by Gopi et al. [7], NNs, specifically an ensemble recurrent neural network (ERNN), are used with the Red Fox Optimization (RFO) technique for crop advice and yield forecasting. Precision farming benefits from increased crop quality and yield with no negative environmental effects. It focuses on monitoring reactions to inter- and intravariability in agricultural systems, management information systems, variable rate technologies, and others. Farmers use conventional farming patterns to make decisions related to crop production and cultivation, but in order to aid in generating data-driven judgments after accounting for variable aspects, an automated crop recommendation system is needed. A crop advice and yield forecasting dataset from the Kaggle repository was used to validate the RFOERNN-CRYP technique experimentally. The technique will help farmers by assisting them in the decision-making process by accounting for the various relevant agricultural parameters. The RFO method is used in the hyperparameter selection procedure, which improves the performance of the ensemble learning process. This also ensures that the accuracy of the model is maintained from the root up. Through the performance of many experimental analyses, it was found that the results of the RFOERNN-CRYP were optimistic, with a maximum R^2 score of 0.9988 and an accuracy of 98.45%. The model proposed in the study can be trialed on real-life, larger datasets in the future. Additionally, the scope of employing deep learning and computer vision technologies-based crop phenotyping detection models will be made possible.

Muthukumaran et al. [8] sought to improve the yield of paddy by utilizing evolutionary computation techniques to predict the nature of the factors influencing growth rate with the intent of improving food security. The researchers focused on analyzing the real-time, day-to-day impact of meteorological phenomena, wind speed, wind direction, and relative humidity instead of relying on historical records. A ROANN Algorithm was used for prediction, and the Multi-Objective Particle Swarm Optimization Algorithm and Genetic Algorithm were utilized to optimize input parameters and reconstruct the database. It is observed that the proposed ROANN algorithm predicts the factors farmers need to focus on while correlating the various meteorological phenomena. The optimization techniques were successful in identifying input parameters from the real-time dataset. Maximum accuracy and minimum error rates were observed upon tuning the NN parameters with hidden layers and learning rates.

A paper by Vignesh et al. [9] aims to predict crop yield using a Visual Geometry Group (VGG) and Discrete Deep Belief Network with the net classification method to create CYP systems to connect raw data to predictions of crop yield. To accomplish this, the researchers gathered data parameters from India's state agriculture website, which served as the foundation for their predictive model. The data parameters collected from the agriculture website were fed into the network, which comprises multiple stacked layers. These layers process the input parameters sequentially, ultimately forming a comprehensive framework for crop production prediction. To optimize the input data for the most effective results, the researchers employed a technique called "tweak chick swarm optimization."

This optimization method refines and preprocesses the input data, ensuring that the most relevant and valuable characteristics are extracted. The resulting refined data is then fed into the classification process as input. To perform the actual classification and predict agricultural yields, the researchers utilized both the Discrete Deep Belief Network and the VGG Net Classifier. The outcome of their approach is notably impressive, surpassing previous methods in terms of predictive accuracy. Specifically, this model achieves an impressive accuracy rate of 97%, underscoring its proficiency in selecting the crop varieties that are likely to yield the highest profits.

9.2.2 Optimization Algorithms

The study by Mohsen Yoosefzadeh et al. [10] aims to improve the genetic yield potential in the soybean crop, which is a major food-grade crop that could potentially be the most sustainable solution to tackling the rising food security concerns and global demand. Five traits were identified first as the most significant soybean yield component attributes were assessed. ML algorithms were used both separately and collectively on the selected data through an ensemble method using the Random Forest (RF) algorithm, Multilayer Perceptron (MLP), and Radial Basis Function (RBF). Out of these techniques, the RBF algorithm was the most efficient with the coefficient of determination (R^2) of 0.81 and the lower mean absolute errors (MAE) of 148.61 kg.$ha^{-1}$and root mean square error (RMSE) values of 185.31 kg.ha^{-1}. The RBF was selected as the subclassifier for the E-B algorithm. Breeding approaches can be modified to improve the genetic prospective of complex traits, such as yield hinged on the secondary attributes that govern the final efficiency and result. Although the results of the study are promising, the measurement of the yield components is labor intensive and, therefore, cannot be applied to large-scale deployments and hence limits the use of this strategy to initiatives for cultivar development. The study will help breeders choose parental lines and carry out fruitful crosses, two crucial processes for breeders looking to develop cultivars with higher genetic production potential.

Predicting yield considering various irrigation kinds is ideal for optimizing resources. CYP plays an integral role in achieving irrigation schedules and assessing labor requirements for storing and reaping. The accuracy of CYP requires a detailed logical understanding of the relativity between yields and collaborative factors. To achieve this connection, large datasets and an efficient model are required. Deep learning has been utilized in this research [11] to assess and analyze complex data like genotypes, weather parameters, maturity groups, etc. An Automated Crop Yield Prediction (ACYP) utilizing the Chaotic Political Optimizer with Deep Learning (CPODL) model is proposed. The steps involved in ACYP-CPODL technique are preprocessing and parameter optimization. For this model, the hyperparameter tuning is by the CPO algorithm. This technique has been able to achieve and produce reasonably successful results with a mean squared error (MSE) of 0.031 and an R^2 score of 0.936. It should, however, be noted that the Bidirectional Long Short-Term Memory (BLSTM) model has also been able to produce results mirroring the aforementioned model to a good extent. A wider range of simulations on benchmark datasets has shown that in a comparative study, the betterment in the predictions made by the ACYP-CPODL model is validated and verified to be true. It has demonstrated success as a technique for estimating agricultural yields. The design of fusion-based prediction models will improve the predictive outcome in the future.

The paper referenced as [12] seeks to harness the extensive data collected by Earth observation (EO) facilities to develop a model that can enhance predictions of crop yields. To achieve this, the study leverages EO data sourced from Sentinal-2 and Landsat-8 satellites to fine-tune the parameters of the APSIM-Maize model. To address the inherent uncertainties associated with soil properties, meteorological variables, and unknown management practices, a Monte-Carlo sampling method was employed. This approach enabled the researchers to account for and propagate these uncertainties effectively. Specifically, using both the mean values and uncertainty measures derived from the soil grids dataset, the study generated 25 alternative representations or ensembles of the soil profile for each site, thus capturing the variability in soil attributes. The research made use of data collected from two key EO satellites, Sentinel-2 and Landsat-8, for calibration purposes. Sentinel-2 provided valuable observations of Leaf Area Index (LAI), while Landsat-8 contributed data related to the Normalized Difference Vegetation Index (NDVI). These specific variables were chosen because previous research has demonstrated their robust predictive power in relation to crop yield. This analysis was instrumental in identifying the parameters within the APSIM model that exert the most significant influence on the prediction of maize LAI. By systematically assessing the sensitivity of these parameters, the research sought to enhance the accuracy and reliability of CYPs within the APSIM-Maize model.

According to the author [13], the production of agriculture involves a considerable deal of uncertainty; thus, it is impossible for any one approach to fully predict the end outcome. This limitation can be overcome by integrating swarm intelligence optimization with the fertilizer effect function to present many possibilities of validation findings under various random settings. This approach is more practical than conventional approaches like regression. It is emphasized that the limitations necessary for this [13] methodology will expand when the algorithm is applied in a real-world setting, perhaps causing results to deviate but not enough to affect the method's application. In order to meet the objectives of scientific management, further work on the approach will concentrate on investigating a collection of predefined fertilization models. It was noted that the algorithm performs well in equation coefficient solving fitting and maximum yield solution. The findings demonstrated an increased fitting degree in the fertilizer impact equation capable of reasonably forecasting the accurate fertilizer application ratio and improving the yield.

The importance of crop production prediction in the agricultural industry is acknowledged in this paper [14]. It intends to make use of well-known ML methods, including long short-term memory and convolutional neural networks. In order to estimate the yield for certain crops, it combines both of those models. The dataset used in this study included information on rainfall from 1901 to 2020, as well as agricultural yield information for cotton, maize, ragi, paddy, and sugarcane. These datasets were gathered from the website of the Indian government. To use all of this data efficiently in this study, it was integrated and preprocessed [15].

The improved optimization function (IOF) reduces the error and converges fast depending upon the back propagation used. After running the models, the following performance metrics were used for evaluation:

- Mean absolute percentage error (MAPE) – 0.836 to 1.245
- MAE – 0.054 to 2.321
- RMSE – 0.071 to 2.321
- Adjusted R-squared – 0.964 to 0.977

9.2.3 Remote Sensing and Imaging

This paper [16] deals with the usage of Unmanned Aerial Systems (UAS) Multispectral Imagery to predict the yield and to optimize the input resource for the corn crop. An unmanned UAS can be used to get images that contain spatial and temporal information, which will be used to develop a statistical model to predict yield based on different phonological stages. Study [16] was conducted at the W. B. Andrews Agriculture Systems Research Farm at Mississippi State, Mississippi, United States. The data used was from the Delta Agricultural Weather Center. The field was subdivided into subplots, each with different treatments to identify which is the most ideal for the corn crop.

This paper realizes the significance of CYP in the agricultural sector. It aims to utilize popular ML techniques such as convolutional neural networks and long short-term memory. It combines both of those models to estimate the yield for certain crops. The data collected for this research were a dataset that contained the rainfall data from 1901 to 2020 and the crop yield data of paddy, maize, ragi, sugarcane, and cotton. These datasets were collected from the Indian government website. All of this data was combined and preprocessed to be used effectively in this study. The paper proposes an enhanced optimized function, i.e., an IOF, to reduce the error in CYP. The IOF function reduces the error and converges fast depending upon the back propagation used. After running the models, the following performance metrics were used for evaluation:

- MAPE – 0.836 to 1.245
- MAE – 0.054 to 2.321
- RMSE – 0.071 to 2.321
- Adjusted R-squared – 0.964 to 0.977

9.2.4 Feature Selection

This paper [17] deals with the aspect of feature selection for optimizing CYPs. Since there are multiple factors which are involved in predicting crop yields, selecting the required ones is important, as multiple factors affect the crop yield differently [17]. Sequential Backward Elimination Feature Selection Algorithm: AH, CL, TK, OW, and Tmax features

Correlation-Based Feature Selection Algorithm: AH, CL, TK, AT, Tmax, SD, NF, PF, and KF

Variance Inflation Factor: AH, CL, TK, TW, OW, RainF, AT, Tmin, Tmax, and SR features

Variance Inflation Factor: AH, CL, TK, TW, OW, RainF, AT, Tmin, Tmax, and R features

RF Variable Importance: AH, TK, OW, NF, PF, KF, and SD

Taking prediction dependent (PD) as the dependent feature and AH, OW, Tmax, TK, and CL as the independent features, we performed multiple linear regression. They have also used an ANN, taking the dependent features as the input layer and PD as the output layer neuron. The final model's accuracy was adjusted using correlation coefficient squared ($R2$) and came out to be 85% when only the selected features were used, whereas when all the features were used, the final accuracy came out to be 84%. This, therefore, could be used to conclude that the forward feature selection algorithm is good for better prediction.

9.2.5 Precision Irrigation

This paper [18] deals with obtaining maximum yield from crops using subsurface water retention technology (SWRT). Here we tackle the process of determining the crop yield using algorithms. We require mainly two different factors to determine the yield. This study [18] was undertaken due to the high priority that sustainable crop cultivation has in our society. We need optimal coordination of food, water, and energy to keep on moving forward. Since water plays a huge role in agricultural development, the most efficient use of water plays a key role in the optimization of crop yield. SWRT is a method developed to improve the soil-water holding capacity in the plant root zone. Using the HYDRUS-2D software, we find the proper membrane design and installation depth in specific soil and weather conditions in sandy soils. Through various tests, it was concluded that the HYDRUS-2D model is highly reliable and has considerable accuracy. Although HYDRUS-2D can predict the water and nutrient accumulation at the root zone of a plant, it cannot simulate crop growth. To compensate for this shortcoming, we deploy the DSSAT software to give insights into crop yield from the specific soil-water mix. Using the proper mathematical modeling and proper use of both software, we can perform crop yield simulation under different SWRT membranes. The tests and results in this paper [18] help us conclude that using both DSSAT and HYDRUS-2D, will help cut down on the human effort of optimizing certain parameters, which will result in increased crop production.

9.3 METHODOLOGIES

9.3.1 Preprocessing

Preprocessing plays a vital role in designing an ML algorithm, as most algorithms require data to learn from and provide predictions. If the data is not up to the mark, the predictions will be really inaccurate.

The preprocessing we have done in the selected dataset includes the following:

1. Outlier Analysis
2. Correlation Analysis
3. Updating Missing Values
4. One-Hot Encoding

Figure 9.1 is the heat map showing the correlation between all variables in our final dataset:

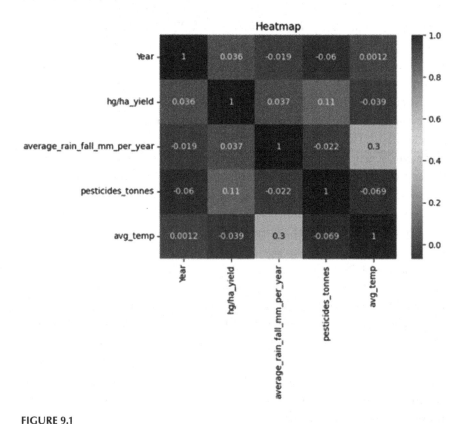

FIGURE 9.1

Heat map correlation for a given set of attributes.

As we can see here, we have taken five variables for analysis, as the others have been eliminated for not having enough relevance for the model to be used. Since the correlation ranges from −1 to 1 in a heat map (−1 indicating strong inverse correlation and 1 indicating strong correlation), ignoring the diagonal in the matrix, the highest change occurs in the crop yield depending on the variable pesticides_tonnes (0.11).

9.3.2 Linear Regression

Linear regression can be defined as a dependent variable's link to one or more independent variables determined statistically. Based on the input value of another variable, it may be used to forecast the value of a variable. In the previously mentioned case, the variable whose value is found is known as the independent variable. The dependent variable is the one that is utilized to estimate the value of the independent variable.

Linear regression aims to find the best-fitting line that can encompass the values predicted and describe the link between the independent variable and dependent variables. It is commonly used to estimate trends in data. It is a productive statistical technique used in ML for predictive analysis. The dependent variable and independent variable are shown to have a linear relationship. Typically, the values of the dependent variables are on the y-axis together with the values of the independent variables. The linear regression model is known as simple linear regression when there is just one input variable.

The best-fit line is calculated using the traditional slope-intercept formula, which is given as follows.

$$Y_i = \alpha_0 + \alpha_1 X_i \tag{9.1}$$

In the Equation 9.1, Y_i denotes the dependant variable(s), which was previously plotted on the y-axis, and X_i denotes the independent variable plotted on the x-axis. The coefficients α_0 and α_1 represent the constant/intercept value and the slope of the best-fit line, respectively.

9.3.3 Polynomial Regression

In certain cases, the points plotted on an x–y regression graph may be nonlinear and, therefore, a single straight best-fit line might not include or account for all the points. To accommodate this scenario, we require the implementation of polynomial regression models. That is, to overcome the conditions of underfitting or overfitting due to the usage of a straight line as the best-fit line, polynomial regression is used. Polynomial regression helps identify the curvilinear association between dependent and independent variables. In polynomial regression, the association between the dependent and independent variables is modeled as a *p*th degree polynomial function.

Only under these conditions can polynomial regression be performed. Additionally, it is expected that the mistakes are random, regularly distributed with a mean of zero, and variance is constant. Some characteristics of polynomial regression are that the degree of the function can be altered to determine the best-fit line for a given set of points, and the model is prone to outliers, which can influence the result. The standard formula for polynomial regression can be given as in Equation 9.2.

$$g(y) = d_0 + d_1 y + d_2 y^2 + \ldots + d_p y^p \tag{9.2}$$

Where *p* is the degree of the polynomial and *d* is the set of coefficients. For lower degrees, the relationship has specific names, depending on the value of when the value of *p* is 2, it is known as a quadratic regression model, and similarly, when the value of *p* is 3 and 4, it is known as cubic and quartic regression models, respectively. The method of least squares is typically used to fit polynomial regression models, minimizing the variance of the coefficients. Selecting the right degree (*p*) for a model is essential as it prevents overfitting or underfitting.

9.3.4 XGBoost

XGBoost stands for eXtreme Gradient Boosting. It is a product of gradient-boosted decision trees that can deliver results with high speeds and performance. It is useful in achieving results on many ML challenges. It largely works on gradient boosting, which is a supervised ensemble learning algorithm, wherein it attempts to predict values by combining the estimates of a set of simpler, weaker models.

It is highly preferred due to its ability to manipulate substantially large datasets and perform well in cases of tasks involving regression and classification. Due to its ability to manage missing values without requiring preprocessing of the input, it is well-liked. Additionally, it includes integrated parallel processing functionality, allowing it to train models quickly with big datasets.

Decision trees are built sequentially in the algorithm. Weight is an important parameter in XGBoost. Weights are associated individually with all the independent variables; these are then inputted into the decision trees, which predict the result. Multiple decision trees are utilized to refine the predicted results by feeding data forwards; this is known as ensemble learning. The first-order derivative (gradient) and second-order derivative (Hessian) of the loss function are calculated for each data point, as depicted in Equations 9.3 and 9.4, respectively. These values are used to build the individual trees and adjust the model.

Regression Gradient (g_i) for MSE:

$$g_i = -2^* \left(y_i - \bar{y}_i \right) \tag{9.3}$$

Regression Hessian (h_i) for MSE:

$$h_i = 2 \tag{9.4}$$

The recursive adjustment process adopted by the algorithm makes it efficient in capturing complex relationships with data, thus making it increasingly accurate at capturing nonlinear patterns and interactions. It uses a regularization value in its objective function. XGBoost includes L1 (Lasso) and L2 (Ridge) regularization terms to control the complexity of the individual trees and prevent overfitting.

Regularized Objective
= Objective + Penalty Regularized Objective
= Objective + Penalty

This is essential in preventing overfitting, which is a condition where the ML model becomes too tailored to the training data that it cannot adapt to a new dataset. In XGBoost, the regularization technique aids in ensuring a strong correlation between the result produced by the model with respect to the training dataset and the test dataset. The ensemble approach adopted also enables the algorithm to harness the strength of multiple models while also mitigating their weaknesses. XGBoost is highly preferred, as it is open-source software that undergoes continuous development by contributing professionals in the field.

9.3.5 KNN

The ML technique K-nearest neighbors, often known as KNN, is frequently employed for classification and regression issues. It is an instance-based algorithm that, unlike other algorithms, does not build an explicit algorithm during its training stage. It is an algorithm based on the supervised learning technique. The KNN algorithm memorizes the entire training dataset and, in response to an unknown data point, locates the K-nearest points in the training dataset and assigns the new value to the class or label based on a majority vote or an average of the k neighbors. This is done by using the Euclidean distance for continuous features and the Hamming distance for categorical features. Euclidean distance between two points *p* and *q* in a d-dimensional space is shown in Equation 9.5:

$$\text{Euclidean Distance}\left(p,q\right) = \sqrt{\sum_{i=1}^{d}\left(P_i - q_i\right)^2} \qquad (9.5)$$

The algorithm has two parameters – namely, *K* and distance metric. *K* is the number of neighbors to consider. It is a crucial hyperparameter in the algorithm, and the accurate selection of the same can significantly impact the performance of the algorithm. Distance metrics are used to find the alikeness between two given data points. While KNN is a simple algorithm, it has some limitations, such as sensitivity to the value of *k*, sensitivity to the choice of distance metric, and computational inefficiency with large datasets.

9.4 RESULTS AND DISCUSSION

This section's primary goal is to present the results of the model experiments. After the various prediction models have been trained, we may use Python 3.8.0 to run simulations. Consistent findings will be produced as the number of tests is increased. As a result, we will assess how effective the ML models outlined are in predicting crop yield overall in this section.

9.4.1 Model Evaluation

- **Mean Squared Error**
  The MSE emerges as a crucial parameter illuminating the predictive efficacy of the ML models employed in this crop production

prediction study. It calculates the squared difference between the anticipated crop yield and the observed yield for each data point in the dataset. The computation of the MSE involves summing up all these squared differences and subsequently taking an average across all the data points. Mathematically, this process can be expressed as in Equation 9.6:

$$\text{MSE} = 1/n * \pounds \left(\text{actual yield} - \text{predicted yield} \right)^2$$

(9.6)

In this context, the variable "n" represents the total count of data points. A lower MSE value indicates that the models – namely, XGBoost, polynomial regression, and KNN – produce predictions that closely correspond to the observed crop yields. A reduced MSE suggests that the models successfully capture the intrinsic patterns and fluctuations within the data, thereby enhancing the overall reliability of their predictions.

As we can see in Figure 9.2, the x-axis represents the name of the model that we've used, and the y-axis represents the respective MSE. The values range from 0.0 to 3.0, as seen on the y-axis. Here, the lower the bar for each model, the better its performance, as the error is lower.

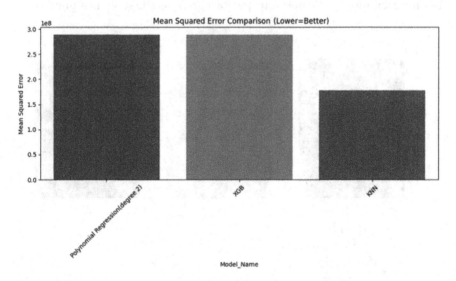

FIGURE 9.2

Mean square error comparisons.

- **R-squared (R^2) Score**

  During the evaluation of how effectively the ML models align with the observed crop yield data, the R-squared (R^2) score assumes a pivotal role. This metric quantifies the proportion of total variance in the crop yield data that the models are capable of accounting for. Utilizing the R^2 score allows for an assessment of the performance of the XGBoost, polynomial regression, and KNN models in relation to a reference model.

  A value of 0 for R^2 (which ranges between 0 and 1) indicates that, much like a basic average model, the models cannot account for any variations in crop yields. Conversely, an R^2 score of 1 signifies that the models are adept at capturing all the variability present in the data, resulting in predictions that precisely match the observed yields.

  In this study, higher R^2 scores demonstrate the models' proficiency in discerning crucial patterns and trends within the crop yield data. These outcomes underscore the importance and reliability of the models' predictions, as they indicate that a substantial portion of the yield variability is accurately captured by the models.

As we can see in Figure 9.3, the x-axis represents the name of the model that we've used, and the y-axis represents the respective R-squared score. The values range from 0.0 to 0.8, as seen on the y-axis. Here, the higher the bar for each model, the better its performance, as the score is higher.

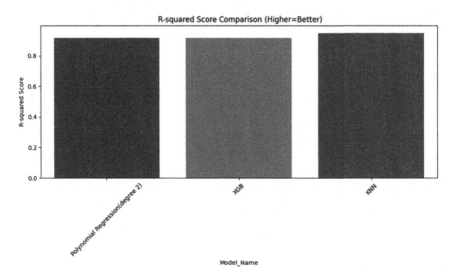

FIGURE 9.3

R-square error comparisons.

9.5 CONCLUSION

ML has become an essential decision-support tool for forecasting crop yields in agriculture, enabling well-informed decisions on crop selection and management all through the growing season. In order to estimate crop yields, this study used a comparative analysis with three ML models: linear regression, polynomial regression, and KNN. To improve the agricultural sector, it is essential to anticipate crop yields based on these variables, which include temperature, pesticide use, rainfall, and climatic conditions. The obtained dataset underwent extensive preprocessing to remove redundant or incorrect data prior to model training. After that, the dataset was split into training and testing subsets so that the performance metrics of each model could be assessed. Running the three ML models we had taken, we got to see that KNN had both the least MSE: $1.776102e + 08$ and the highest R^2 score: 0.948888. Although it performed well in these metrics, KNN should be used carefully, as it can tend to overfit the data, and if not, XGBoost will provide a similar result in most cases. These models can be used to improve the efficiency of farming by helping provide efficient use of materials and resources. This study highlights how ML can improve agricultural productivity and decision-making, thereby assisting in the farming industry's sustainable growth. In order to optimize crop output and adapt to shifting environmental conditions, farmers and policymakers can use ML to their advantage. This will ensure food security and economic growth in the agricultural sector.

REFERENCES

1. "Economic Survey 2022–23: Agri growth rate falls but sector still resilient amid pandemic shock," www.downtoearth.org.in. https://www.downtoearth.org.in/news/agriculture/economic-survey-2022-23-agri-growth-rate-falls-but-sector-still-resilient-amid-pandemic-shock-87406
2. N. Alex, C. C. Sobin, and J. Ali, "A comprehensive study on smart agriculture applications in India," *Wireless Personal Communications*, vol. 129, no. 4, pp. 2345–2385, Mar. 2023. doi: 10.1007/s11277-023-10234-5
3. K. P. Kumar, A. Unal, V. J. Pillai, H. Murthy, and M. Niranjanamurthy, eds., *Data Engineering and Data Science: Concepts and Applications*. 2023. ISBN: 9781119841876.
4. A. Nigam, S. Garg, A. Agrawal, and P. Agrawal, "Crop yield prediction using machine learning algorithms," *2019 Fifth International Conference on Image Information Processing (ICIIP)*, Shimla, India, 2019, pp. 125–130. doi: 10.1109/ICIIP47207.2019.8985951

5. K. P. Kumar, H. Murthy, V. J. Pillai, and B. R. Prathap, "Blockchain-enabled model for minimizing post harvest losses," *ECS Transactions*, vol. 107, no. 1, p. 17475, 2022. doi: 10.1149/10701.17475ecst

6. I. A. Supro, "Rice yield prediction and optimization using association rules and neural network methods to enhance agribusiness," *Indian Journal of Science and Technology*, vol. 13, no. 13, pp. 1367–1379, Apr. 2020. doi: 10.17485/ijst/v13i13.79

7. P. S. S. Gopi and M. Karthikeyan, "Red fox optimization with ensemble recurrent neural network for crop recommendation and yield prediction model," *Multimedia Tools and Applications*, vol. 83, pp. 1–21, Jul. 2023.

8. S. Muthukumaran, P. Geetha, and E. Ramaraj, "Multi-objective optimization with artificial neural network based robust paddy yield prediction model," *Intelligent Automation & Soft Computing*, vol. 35, no. 1, pp. 215–230, 2023. doi: 10.32604/iasc.2023.027449

9. K. Vignesh, A. Askarunisa, and A. M. Abirami, "Optimized deep learning methods for crop yield prediction," *Computer Systems Science and Engineering*, vol. 44, no. 2, pp. 1051–1067, 2023. doi: 10.32604/csse.2023.024475

10. M. Yoosefzadeh-Najafabadi, D. Tulpan, and M. Eskandari, "Application of machine learning and genetic optimization algorithms for modeling and optimizing soybean yield using its component traits," *PLOS ONE*, vol. 16, no. 4, p. e0250665, Apr. 2021. doi: 10.1371/journal.pone.0250665

11. G. Sunitha et al., "Modeling of chaotic political optimizer for crop yield prediction," *Intelligent Automation & Soft Computing*, vol. 34, no. 1, pp. 423–437, 2022. doi: 10.32604/iasc.2022.024757

12. H. Dokoohaki, T. Rai, M. Kivi, P. Lewis, J. Gómez-Dans, and F. Yin, "Linking remote sensing with APSIM through emulation and Bayesian optimization to improve yield prediction," *Remote Sensing*, vol. 14, no. 21, pp. 5389, Oct. 2022. doi: 10.3390/rs14215389

13. C. Chen, X. Wang, H. Chen, C. Wu, M. Mafarja, and H. Turabieh, "Towards precision fertilization: Multi-strategy grey wolf optimizer based model evaluation and yield estimation," *Electronics*, vol. 10, no. 18, p. 2183, Sep. 2021. doi: 10.3390/electronics10182183

14. U. Bhimavarapu, G. Battineni, and N. Chintalapudi, "Improved optimization algorithm in LSTM to predict crop yield," *Computers*, vol. 12, no. 1, p. 10, Jan. 2023. doi: 10.3390/computers12010010

15. K. P. Kumar, H. Murthy, V. J. Pillai, B. R. Prathap, M. Moses, and Y. Urs, "Data analysis and machine learning observation on production losses in the food processing industry," *2023 IEEE International Conference on Contemporary Computing and Communications (InC4)*, Bangalore, India, 2023, pp. 1–4. doi: 10.1109/InC457730.2023.10263147

16. R. Barzin, R. Pathak, H. Lotfi, J. J. Varco, and G. C. Bora, "Use of UAS multispectral imagery at different physiological stages for yield prediction and input resource optimization in corn," *Remote Sensing*, vol. 12, no. 15, pp. 2392, Jul. 2020. doi: 10.3390/rs12152392

17. P. S. Maya Gopal and R. Bhargavi, "Selection of important features for optimizing crop yield prediction," *International Journal of Agricultural and Environmental Information Systems*, vol. 10, no. 3, pp. 54–71, Jul. 2019. doi: 10.4018/ijaeis.2019070104

18. P. C. Roy, A. Guber, M. Abouali, A. P. Nejadhashemi, K. Deb, and A. J. M. Smucker, "Crop yield simulation optimization using precision irrigation and subsurface water retention technology," *Environmental Modelling & Software*, vol. 119, pp. 433–444, Sep. 2019. doi: 10.1016/j.envsoft.2019.07.006

10

Fundamentals of AI and Machine Learning with Specific Examples of Application in Agriculture

Manoj Kumar Mahto, P. Laxmikanth,
and V.S.S.P.L.N. Balaji Lanka
Vignan Institute of Technology and Science, Hyderabad, India

10.1 INTRODUCTION

The integration of artificial intelligence (AI) and machine learning (ML) techniques has ushered in a paradigm shift across diverse sectors, propelling innovative solutions that redefine conventional practices. Within this transformative landscape, the agricultural domain stands prominently, benefiting from the power of AI and ML to address complex challenges and elevate traditional farming practices. This chapter delves into the core fundamentals of AI and ML, contextualizing their principles within the realm of agriculture. By navigating the theoretical foundations and practical applications of these technologies, readers will gain a comprehensive understanding of how AI and ML are reshaping the agricultural landscape. AI and ML, once confined to the realm of science fiction, have now become pivotal tools in real-world applications. In agriculture, where the intricate interplay of weather patterns, soil conditions, and crop health shapes outcomes, AI and ML offer a powerful synergy. This chapter unravels the foundational concepts of AI and ML, providing insights into their inner workings. AI, often referred to as the simulation of human intelligence processes, empowers computers to execute tasks that would otherwise demand human intelligence. ML, a subset of AI, equips computers with the ability to learn from data and progressively improve their performance. Both technologies thrive on data, and in agriculture, where

DOI: 10.1201/9781003485179-10

data-rich variables define outcomes, this symbiotic relationship takes center stage. A pivotal aspect of AI and ML in agriculture is their ability to harness data's potential to make informed decisions. Data collection and preprocessing constitute the bedrock, ensuring that the data fed into AI and ML models is accurate, reliable, and relevant. This process involves gathering diverse datasets related to weather patterns, soil attributes, crop characteristics, and historical yields. Subsequently, feature extraction and engineering highlight the art of selecting the most relevant attributes from the data and refining it into a usable form that optimizes model performance. Model selection and evaluation further underscore the significance of informed choices. The process entails choosing suitable algorithms that align with the intricacies of the problem at hand. These algorithms undergo rigorous evaluation, ensuring that the model's predictions are dependable and actionable. The fusion of these stages results in AI and ML models that are primed to revolutionize agriculture.

The applications of AI and ML in agriculture span a wide range of challenges and opportunities. Key areas include crop yield prediction, which empowers farmers to optimize resource allocation and productivity, and disease detection, which facilitates rapid diagnosis and intervention. Precision farming, coupled with decision support systems, enhances efficiency through intelligent practices. Real-world examples highlight their impact, such as crop disease detection using convolutional neural networks (CNN), precision agriculture with Decision Trees and Internet of Things (IoT) devices, and yield prediction via regression algorithms. These instances underscore AI and ML's transformation from theory to practical change agents in agriculture. However, ethical concerns related to data privacy and algorithmic transparency arise, necessitating thoughtful consideration. Additionally, ensuring equitable access and benefits across all agricultural stakeholders is crucial, given the technologies' profound transformative potential.

10.2 FUNDAMENTALS OF AI AND ML

The fundamentals of AI and ML encompass key concepts that underpin their applications and advancements in various industries, including agriculture. AI refers to the creation of intelligent systems that can simulate human-like behaviors, while ML involves the development of algorithms that enable machines to learn from data and improve their performance over time.

10.2.1 AI

AI encompasses a range of techniques and approaches aimed at creating machines or systems that exhibit human-like intelligence. These systems can perform tasks such as problem-solving, decision-making, understanding natural language, and recognizing patterns. AI techniques include rule-based systems, expert systems, neural networks, and natural language processing (NLP).

10.2.1.1 AI in Agriculture

AI in agriculture is spearheading a transformative shift in the way we cultivate crops. With the integration of cutting-edge technologies like agriculture robots and intelligent spraying systems, farmers can now precisely manage their fields, optimizing crop yields while reducing resource consumption. These AI-driven solutions offer real-time crop and soil monitoring, enabling proactive responses to changing conditions and facilitating predictive insights into weather patterns and pest outbreaks. Additionally, AI plays a pivotal role in disease diagnosis, swiftly identifying and addressing potential threats to crop health. Moreover, AI-driven price forecasts provide farmers with valuable information for strategic decision-making. In essence, AI is not just revolutionizing agriculture; it's paving the way for a more sustainable, efficient, and economically viable future for the industry. Figure 10.1 illustrates the comprehensive application of AI in the field of agriculture.

10.2.2 ML

ML is a subset of AI that focuses on developing algorithms that allow computers to learn from data and make predictions or decisions based on that learning. ML algorithms are categorized into supervised, unsupervised, and reinforcement learning, each catering to different types of tasks. Supervised learning involves training the model on labeled data, unsupervised learning involves finding patterns in unlabeled data, and reinforcement learning involves training agents to take actions to maximize rewards.

10.2.2.1 ML in Agriculture

ML is revolutionizing agriculture by optimizing various facets of the industry, including crop management, water management, soil management,

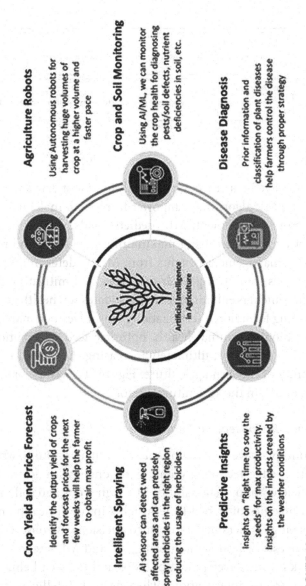

FIGURE 10.1

AI in agriculture.

Crop Management
• This category involves studies concerning:
• Yields Prediction, Disease Detection, Weed Detection, Crop Recognition, and Crop Quality

Water Management
• This category is associated with the optimal use of water resources

Soil Management
• This category is related to soil protection and soil management aspects

Livestock Management
• This category includes the management pertaining to: Animal Welfare and Livestock Production

FIGURE 10.2
ML in agriculture.

and livestock management. Through the analysis of vast datasets and real-time sensor data, ML algorithms empower farmers to make data-driven decisions. In crop management, ML predicts yields, disease outbreaks, and ideal planting times, ensuring maximum productivity and resource efficiency. Water management benefits from ML by determining precise irrigation schedules, conserving water resources, and mitigating drought risks. Soil management is enhanced, as ML evaluates soil health and nutrient levels, allowing for tailored fertilization plans. Livestock management utilizes ML to monitor animal health, optimize feed distribution, and enhance breeding programs, ultimately increasing the overall efficiency and sustainability of modern agriculture. Figure 10.2 visually represents the utilization of ML in the agricultural sector.

10.2.2.2 AI and ML Concepts in Agriculture

AI and ML concepts have ushered in a new era in agriculture, where the fusion of cutting-edge technology and traditional farming practices is optimizing efficiency and sustainability. Through crops and soil monitoring, these technologies offer real-time insights into soil health and crop conditions, helping farmers make informed decisions. By observing crop maturity and livestock health monitoring, AI and ML systems provide timely feedback on the development of crops and the well-being of livestock, enabling precise interventions when needed. Intelligent spraying systems and automatic feeding mechanisms further enhance precision agriculture, reducing resource wastage and ensuring the judicious use of fertilizers and pesticides. In essence, these concepts are revolutionizing the industry by fostering a more informed, efficient, and eco-conscious

Intelligent Spraying
• UAVs with computer vision AI to automate spraying pesticides and fertilizers uniformly

Automatic Weeding
• Weeding robots to navigate and destroy existing weeds without applying herbicide

Livestock Health Monitoring
• Tracking and explaining cattle through Cattle Eye technology training data

Crops and Soil Monitoring
• Tracking crop health and predicting yields through visual sensing

Observing Crop Maturity
• Detecting maturity of vegetables using drones or unmanned aerials vehicles

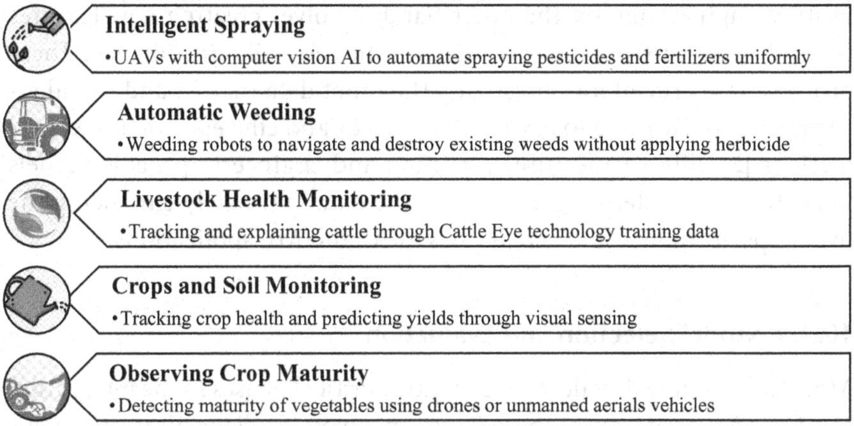

FIGURE 10.3
AI and ML concepts in agriculture.

approach to farming. Figure 10.3 presents an overview of the concepts of AI and ML in the context of agriculture.

10.2.3 Data Collection and Preprocessing

Data collection and preprocessing constitute fundamental steps in the integration of AI and ML techniques. The process involves gathering diverse datasets related to specific domains, ensuring their accuracy, completeness, and relevance for subsequent analysis. Data preprocessing, as a critical precursor to analysis, encompasses activities like data cleaning, transformation, and feature extraction to refine raw data into a usable form. These stages are pivotal for enhancing the quality of input data, which directly impacts the performance and reliability of AI and ML models [1]. By focusing on data quality and preparation, organizations can ensure that the subsequent AI and ML processes are based on accurate and meaningful information.

10.2.4 Feature Extraction and Engineering

Feature extraction and engineering play a central role in harnessing the power of AI and ML techniques. These processes involve identifying and selecting relevant attributes or features from the raw data that contribute to the predictive capabilities of models. Feature extraction focuses on transforming the data into a format that is more suitable for analysis, often involving techniques like dimensionality reduction or transformation.

Feature engineering, on the other hand, involves creating new features or modifying existing ones to enhance the model's performance. These processes are crucial for optimizing the input data that AI and ML algorithms rely on, leading to improved accuracy and efficiency in model predictions [2]. Effective feature extraction and engineering enable models to capture the underlying patterns in the data, ultimately enhancing the overall performance and capabilities of AI and ML applications.

10.2.5 Model Selection and Evaluation

Model selection and evaluation constitute critical phases in the integration of AI and ML techniques. These stages are pivotal in determining the algorithms that best suit the problem at hand and assessing their performance. Model selection involves choosing from a range of algorithms based on their suitability for the specific task, data characteristics, and objectives. Subsequently, model evaluation involves testing the chosen algorithms using validation and testing datasets, measuring their accuracy, precision, recall, and other relevant metrics to assess their effectiveness [3]. Effective model selection and evaluation are imperative for ensuring that AI and ML models produce reliable predictions and insights, guiding decision-making processes in various domains.

10.3 APPLICATIONS OF AI AND ML IN AGRICULTURE

The applications of AI and ML in agriculture have revolutionized traditional farming practices by leveraging data-driven insights to address complex challenges. These technologies have the potential to transform various aspects of agriculture, from improving crop yield predictions to enhancing disease detection and management. For instance, AI-powered models can analyze historical weather data, soil characteristics, and other variables to provide accurate predictions of crop yields, enabling farmers to make informed decisions about resource allocation and planning [4]. Additionally, AI and ML algorithms can detect and diagnose crop diseases by analyzing images of plants, leaves, or fruits, enabling timely interventions to mitigate losses and optimize crop health [5]. The integration of AI and ML into precision agriculture has paved the way for optimized resource utilization through targeted irrigation and fertilization, leading to increased efficiency and sustainability. These applications underscore the transformative potential of AI and ML in revolutionizing the agricultural sector. Figure 10.4 visually illustrates the various applications of AI and ML within the agricultural domain.

10.3.1 Crop Yield Prediction

Crop yield prediction, a cornerstone application within the realm of AI and ML in agriculture, has gained substantial attention due to its potential to optimize resource allocation and enhance food security. This predictive approach relies on the integration of historical and real-time data, encompassing factors such as weather patterns, soil conditions, and agricultural practices. By employing advanced regression and ensemble techniques, AI and ML models can decipher intricate relationships between these variables and predict future crop yields with remarkable accuracy [6]. Such predictions provide invaluable insights for farmers, allowing them to make informed decisions regarding planting strategies, irrigation scheduling, and harvest planning. Ultimately, the integration of AI and ML into crop yield prediction not only improves agricultural efficiency but also contributes to sustainable practices by minimizing resource wastage and supporting global food supply chains. Figure 10.5 depicts the process of crop yield prediction.

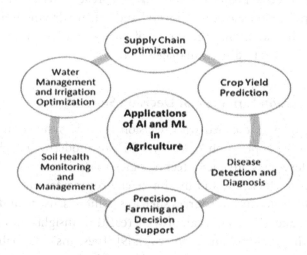

FIGURE 10.4

Application of AI and ML in agriculture.

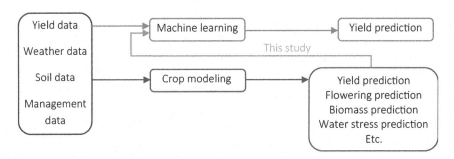

FIGURE 10.5
Crop yield prediction.

10.3.2 Disease Detection and Diagnosis

Disease detection and diagnosis in agriculture have been significantly transformed by the integration of AI and ML techniques. This application plays a pivotal role in the early identification and management of crop diseases, mitigating yield losses, and ensuring food security. By utilizing image recognition and classification algorithms, AI and ML models can analyze images of plants, leaves, or fruits to accurately identify and diagnose various diseases [7]. These models learn from large datasets of images, allowing them to distinguish subtle visual cues that human eyes might miss. With timely and accurate disease detection, farmers can implement targeted interventions such as precise pesticide application or isolation of infected plants, minimizing the spread of diseases across crops [8]. This application showcases how AI and ML are driving innovation in agriculture by harnessing the power of data and automation to address critical challenges in disease management.

10.3.3 Precision Farming and Decision Support

Precision farming, coupled with decision support systems, represents a transformative application of AI and ML in modern agriculture. This approach leverages advanced technologies to optimize resource allocation, enhance efficiency, and minimize environmental impacts. By integrating data from various sources such as satellites, sensors, and historical records, AI and ML models can provide real-time insights into soil conditions, weather patterns, and crop health [9]. These insights enable farmers to make informed decisions about when and where to apply irrigation, fertilizers, and pesticides, tailoring interventions to specific areas of need.

This not only increases yields but also reduces resource wastage and environmental pollution. Additionally, decision support systems equipped with AI algorithms can generate recommendations based on complex data analysis, aiding farmers in making timely and strategic choices for optimal crop management. The amalgamation of precision farming and decision support systems highlights how AI and ML are revolutionizing traditional agricultural practices by harnessing data-driven intelligence for sustainable and efficient outcomes.

10.3.4 Soil Health Monitoring and Management

Soil health monitoring and management have witnessed a paradigm shift with the integration of AI and ML techniques. This application offers a data-driven approach to assessing soil quality, nutrient content, and erosion risks, providing valuable insights for sustainable land management. By incorporating sensor data, historical records, and environmental variables, AI and ML models can analyze complex relationships within the soil ecosystem [10]. These models can identify trends and patterns that enable farmers to make informed decisions about soil amendments, irrigation schedules, and erosion control strategies. This not only enhances crop productivity but also contributes to long-term soil conservation and improved agricultural sustainability. The utilization of AI and ML in soil health monitoring exemplifies how technological innovation can empower farmers to make precision-driven decisions that optimize land use and resource utilization.

10.3.5 Water Management and Irrigation Optimization

Water management and irrigation optimization have been revolutionized by the integration of AI and ML techniques. This application addresses the critical need for efficient water usage in agriculture, especially in water-scarce regions. By leveraging data from sensors, weather forecasts, and soil moisture levels, AI and ML models can dynamically adjust irrigation schedules and volumes to match the specific water requirements of crops [11]. These models learn from historical and real-time data to make precise predictions, minimizing water wastage and enhancing crop yield. Moreover, AI-driven irrigation systems can respond to changing conditions, ensuring that crops receive the optimal amount of water without overirrigation. This not only conserves water resources but also reduces energy consumption associated with pumping and distribution. The integration of AI and ML in water management underscores their potential to address sustainability challenges in agriculture while ensuring productive and resilient farming systems.

10.4 CASE STUDIES

Case studies serve as compelling exemplars of the practical application and impact of AI and ML in diverse sectors, including agriculture. These real-world instances provide tangible evidence of how AI and ML technologies can drive transformative change. Case studies often feature innovative solutions that address specific challenges, such as crop disease detection, yield prediction, and precision farming. Through the integration of AI and ML algorithms, these studies showcase enhanced accuracy, efficiency, and sustainability in agricultural practices [12]. By analyzing these cases, stakeholders can glean insights into the potential benefits, challenges, and strategies for the successful implementation of AI and ML technologies in the agricultural domain, thereby fostering informed decision-making and fostering a resilient food system.

10.4.1 Crop Disease Detection Using CNN

Crop disease detection using CNNs has emerged as a prominent application of AI and ML in agriculture. CNNs are specialized deep-learning models designed to process visual data, making them well-suited for analyzing images of crops and detecting diseases. By learning from large datasets of annotated images, CNNs can identify subtle visual patterns associated with various diseases, enabling accurate and early detection [13]. These models have showcased impressive performance in classifying diseases across different crops, offering potential solutions for timely interventions and targeted management. The utilization of CNNs in crop disease detection showcases the capability of AI and ML to contribute to sustainable agriculture by enhancing disease management practices. Figure 10.6 showcases the implementation of crop disease detection through the utilization of CNNs.

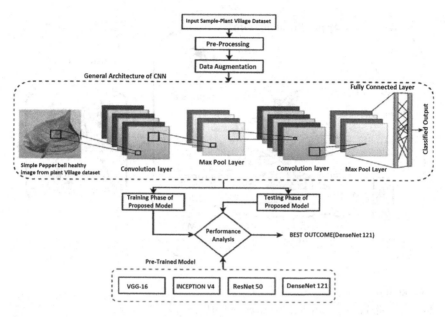

FIGURE 10.6

Crop disease detection using CNNs.

10.4.2 Precision Agriculture with Decision Trees and IoT

Precision agriculture, facilitated by Decision Trees and the IoT, stands as a prime illustration of the transformative impact of AI and ML in the agricultural landscape. Decision Trees are ML algorithms that excel in classification and regression tasks, making them valuable tools for guiding agricultural decisions. When integrated with IoT devices that collect real-time data from fields, Decision Trees can provide actionable insights into crop health, soil moisture, and environmental conditions [14]. This synergy allows farmers to implement precise interventions, such as targeted irrigation and nutrient application, optimizing resource utilization and enhancing yield. The fusion of Decision Trees and IoT underscores the potential of AI and ML to revolutionize traditional farming practices by enabling data-driven precision in agricultural management. Figure 10.7 illustrates the implementation of precision agriculture, integrating Decision Trees and the IoT for improved farming practices.

10.4.3 Yield Prediction Using Regression Algorithms

Yield prediction using regression algorithms exemplifies the application of AI and ML in agriculture, offering valuable insights into future crop

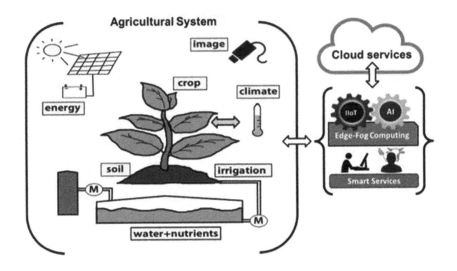

FIGURE 10.7

Precision agriculture with decision trees and IoT.

production. Regression algorithms, a subset of ML techniques, enable the modeling of relationships between input variables and target outcomes. By analyzing historical data encompassing factors such as weather conditions, soil properties, and agricultural practices, these algorithms can generate predictive models that estimate crop yields with accuracy [15]. Such predictions empower farmers with data-driven insights for informed decision-making, including resource allocation and market planning. The successful application of regression algorithms in yield prediction showcases the potential of AI and ML to enhance agricultural productivity and resilience. Figure 10.8 visualizes the process of yield prediction utilizing regression algorithms.

10.5 CHALLENGES AND FUTURE DIRECTIONS

The integration of AI and ML concepts in agriculture presents both opportunities and challenges that warrant consideration. While AI and ML offer transformative potential, several challenges must be addressed to ensure their effective implementation and sustainable impact in agriculture. Data quality and availability remain a significant concern, as reliable and comprehensive datasets are essential for training accurate models [16].

FIGURE 10.8

Yield prediction using regression algorithms.

Privacy and security issues also come into play, particularly as agricultural data often involves sensitive information about farming practices and land management. Ensuring the ethical and responsible use of AI and ML technologies is paramount to building trust among stakeholders [4].

10.5.1 Challenges

Furthermore, the interpretability and explain ability of AI and ML models pose challenges in gaining insights into how decisions are made. Black-box models may hinder the adoption of these technologies, especially in sectors like agriculture, where transparency is crucial [17]. Bridging the gap between technical experts and domain-specific farmers is another challenge. Effective communication and education are necessary to empower farmers to understand, trust, and utilize AI-powered solutions [7].

10.5.1.1 Data Quality and Availability

The necessity of reliable and comprehensive datasets for training accurate AI and ML models in the field of agriculture cannot be overstated. In this context, the quality and composition of the data have a direct impact on the performance and robustness of the models. Inconsistent, incomplete, or biased data can significantly hinder the effectiveness of predictive models and decision-making processes [18]. The accuracy of predictions and the reliability of insights depend on the integrity of the input data, making data quality assurance a critical concern. Ensuring that datasets encompass a diverse range of conditions, scenarios, and variables is essential to avoid biased or skewed model outcomes. Addressing these data-related challenges is pivotal to unlocking the full potential of AI and ML in improving agricultural practices and outcomes.

10.5.1.2 Privacy and Security

Agricultural data frequently encompasses a wealth of sensitive information related to farming practices, land management strategies, and even personal details of farmers. This wealth of personal and proprietary data underscores the critical importance of ensuring robust data privacy and security mechanisms within the context of AI and ML applications in agriculture. As such, safeguarding this data becomes paramount not only to comply with regulatory requirements but also to foster trust and cooperation among farmers, stakeholders, and technology developers [19]. Effective data encryption, access controls, and anonymization techniques are essential to protect confidential information from unauthorized access or breaches. By prioritizing data privacy and security, the agricultural industry can create an environment where stakeholders are confident in the responsible use of AI and ML technologies for enhancing productivity and sustainability.

10.5.1.3 Interpretability and Explainability

The opaqueness inherent in black-box AI and ML models can significantly impede the comprehension of decision-making mechanisms, a concern that holds even more significance within the agricultural domain where transparency is of paramount importance [17]. The inability to discern the rationale behind predictions and recommendations hinders the adoption of these technologies among stakeholders. Given the complex and high-stakes nature of agricultural practices, the need for interpretable models becomes increasingly evident. To establish trust and foster wider acceptance, efforts are directed toward developing AI and ML models with enhanced interpretability. By employing techniques that allow models to provide explanations for their decisions, stakeholders gain insights into the factors influencing outcomes. Such transparent models not only facilitate informed decision-making but also aid in identifying potential biases, errors, or shortcomings, contributing to responsible AI deployment in agriculture.

10.5.1.4 Communication and Education

Narrowing the divide between technical experts engaged in AI solution development and the end-users, who are often farmers, presents a formidable challenge. This challenge is rooted in the distinct backgrounds, terminologies, and perspectives of these two groups. Effective communication and education emerge as pivotal components to bridge this gap and facilitate the successful adoption of AI-powered solutions in agriculture [7]. By translating complex technical concepts into accessible language and relatable scenarios, experts can convey the value and benefits of AI to farmers. Additionally, educational initiatives can empower farmers with the knowledge required to comprehend the potential applications, limitations, and implications of AI in their practices. Such initiatives foster an environment where farmers can make informed decisions, trust the technology, and seamlessly integrate AI-powered solutions into their agricultural processes.

10.5.2 Future Directions

As for future directions, the refinement of AI and ML algorithms tailored specifically for agricultural challenges holds promise. Developments in Explainable AI (XAI) could address the transparency concern, providing more interpretable models that enable stakeholders to comprehend decision-making processes. Integrating AI with other technologies, such as IoT devices, remote sensing, and robotics, can further enhance data collection and decision support systems. Collaborative efforts involving researchers, agricultural experts, and policymakers are essential to create frameworks that encourage responsible AI deployment, address challenges, and promote equitable access to AI benefits. Figure 10.9 outlines the potential future directions in the field of agriculture.

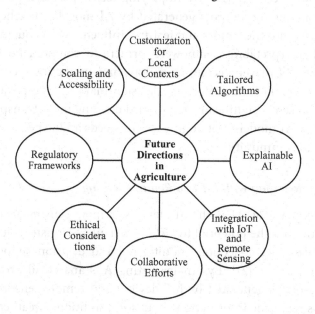

FIGURE 10.9
Future directions in agriculture.

10.5.2.1 Tailored Algorithms

Tailoring AI and ML algorithms to effectively address the unique challenges encountered in agriculture has the potential to significantly amplify the accuracy and applicability of these technologies in farming contexts. This approach involves understanding the intricacies of agricultural systems, such as crop growth patterns, soil characteristics, weather dynamics, and pest behavior, and designing algorithms that leverage this domain-specific knowledge [20]. By customizing algorithms to align with the nuances of agriculture, the resulting models can provide more relevant and actionable insights to farmers. This adaptability not only enhances the predictive power of AI but also increases its practical utility in real-world agricultural scenarios.

10.5.2.2 Explainable AI

Progress in XAI has the potential to drive the development of models that are inherently more interpretable, offering a solution to the challenge of understanding the reasoning behind AI-based decisions. This progress becomes particularly valuable in agriculture, where transparent decision-making processes are crucial for building trust among farmers, stakeholders, and end-users [17]. By employing techniques that provide clear explanations for the outputs generated by AI models, stakeholders can gain insights into the factors and features influencing the outcomes. This enhanced interpretability empowers farmers to comprehend the rationale behind AI-driven recommendations and predictions, contributing to their acceptance and informed decision-making. The advancements in XAI represent a significant step toward creating more transparent and trustworthy AI models that align with the needs and expectations of the agricultural community.

10.5.2.3 Integration with IoT and Remote Sensing

The convergence of AI with IoT devices, remote sensing technologies, and robotics has the potential to yield powerful synergistic solutions that revolutionize data collection, monitoring, and decision support in the agricultural sector [21]. By amalgamating AI's analytical prowess with the data streams generated by IoT devices and remote sensors, farmers can access real-time information about soil conditions, weather patterns, crop health, and machinery operations [22]. This dynamic data ecosystem empowers farmers to make timely and informed decisions, enabling proactive interventions to address challenges like irrigation management, pest outbreaks, and yield optimization. Moreover, the integration of robotics further amplifies these capabilities, facilitating precise and automated actions such as targeted pesticide application or crop harvesting [23]. This holistic approach capitalizes on AI's ability to process complex data, creating a seamless fusion of technologies that augment agricultural productivity, sustainability, and resilience.

10.5.2.4 Collaborative Efforts

Effective collaboration among researchers, agricultural experts, policymakers, and technology developers emerges as a fundamental prerequisite for the successful integration of AI in agriculture. This collaborative approach is imperative due to the multidimensional nature of challenges and opportunities presented by AI deployment [24]. By fostering a cross-disciplinary dialogue, stakeholders can collectively formulate frameworks that not only facilitate the responsible and ethical deployment of AI but also address the intricacies associated with data privacy, bias mitigation, and equitable access to technological benefits. These collaborative efforts culminate in the development of guidelines and regulations that promote transparent and fair AI adoption, ultimately contributing to the creation of an inclusive and sustainable agricultural ecosystem.

10.5.2.5 Customization for Local Contexts

Tailoring AI solutions to align with specific local agricultural practices, diverse climate conditions, and intricate socio-economic factors is imperative for harnessing the full potential of these technologies and ensuring their efficacy. Agricultural systems vary widely across regions, and a one-size-fits-all approach may not yield optimal outcomes [25]. By customizing AI solutions to accommodate local nuances, such as crop cycles, soil types, and resource availability, the technology can be optimized to yield actionable insights that resonate with the unique needs of farmers. Moreover, accounting for socio-economic factors ensures that the adoption of AI solutions is both relevant and beneficial to local communities, fostering sustained engagement and participation. This adaptability enhances the overall effectiveness of AI applications in agriculture, driving positive outcomes that extend beyond productivity gains.

10.5.2.6 Ethical Considerations

Dealing with ethical considerations related to the implementation of AI in agriculture assumes critical significance, encompassing imperatives such as equitable technology access, bias mitigation, and safeguarding the welfare of farm workers. These concerns underscore the need for a comprehensive and responsible approach to AI deployment in the agricultural sector. Ensuring equitable access to AI technology becomes pivotal to preventing the digital divide and enabling all farmers, irrespective of their resources or location, to benefit from technological advancements [26]. The crucial aspect of minimizing biases in AI systems is vital to prevent discriminatory outcomes and ensure fair decision-making, particularly in processes like crop yield prediction or disease diagnosis. Lastly, preserving the well-being of farm workers is paramount, as automation and AI integration can reshape labor dynamics, necessitating thoughtful planning and measures to prevent any adverse impacts on livelihoods [27]. Adhering to these ethical considerations is indispensable for fostering a sustainable and equitable AI transformation in agriculture.

10.5.2.7 Regulatory Frameworks

Formulating suitable regulations and policies for the utilization of AI applications in the agricultural sector holds significant potential to establish a framework that guides the safe and responsible integration of these technologies. The multifaceted nature of AI's impact necessitates governance that encompasses aspects such as data privacy, security, ethical considerations, and compliance with industry standards [28]. By devising regulations that address these concerns, policymakers can facilitate the ethical deployment of AI while maintaining a fair and level playing field for all stakeholders involved. These regulations serve as a safeguard against potential risks and help build a conducive environment where AI adoption is characterized by transparency, accountability, and adherence to best practices.

10.5.2.8 Scaling and Accessibility

Ensuring the accessibility of AI and ML solutions to small-scale farmers and remote regions stands as a pivotal imperative in achieving broad-spectrum impact and diminishing disparities in the adoption of agricultural technology. In many regions, small-scale farmers form the backbone of the agricultural sector, and extending the benefits of AI and

ML to them is crucial for inclusive growth [29]. Additionally, reaching remote areas with limited connectivity and resources requires innovative solutions to bridge the digital divide and bring the advantages of AI to those who need them most. By prioritizing accessibility, policymakers, technology developers, and stakeholders contribute to the democratization of agricultural advancements, empowering farmers across different scales and geographical contexts.

While AI and ML have shown great potential in transforming agriculture, several challenges need to be navigated to ensure their successful integration. Addressing data quality, privacy, ethical considerations, and interpretability are crucial steps. Looking ahead, refining algorithms, embracing XAI, and fostering collaborative ecosystems can drive AI's role in reshaping agriculture for a more sustainable and resilient future.

10.6 CONCLUSION

The fusion of AI and ML with agriculture holds immense promise, offering transformative potential across various aspects of the sector. This comprehensive exploration has shed light on the fundamental concepts of AI and ML, their data-driven nature, feature extraction, model selection, and evaluation. The applications discussed, ranging from crop yield prediction to disease detection and precision farming, underscore how these technologies are reshaping traditional practices to enhance productivity, sustainability, and resource optimization. Moreover, the challenges and considerations surrounding data quality, privacy, transparency, and stakeholder collaboration emphasize the need for responsible and ethical deployment.

As AI-driven solutions continue to advance, the integration of XAI and collaboration between experts and farmers are expected to enhance transparency and adoption. The evolving landscape of AI in agriculture necessitates the development of frameworks that address ethical, regulatory, and accessibility concerns. These efforts are vital to ensure that AI and ML technologies are not only accessible but also contribute to bridging disparities in technology adoption among different agricultural stakeholders. Ultimately, this integration presents an exciting trajectory toward a more resilient, productive, and sustainable agricultural future, where technology augments human expertise for the collective benefit of farmers, communities, and the global food supply.

REFERENCES

[1] H. Witten, E. Frank, and M. A. Hall, *Data Mining: Practical Machine Learning Tools and Techniques* (4th ed.), Morgan Kaufmann, 2016.

[2] I. Guyon and A. Elisseeff, "An Introduction to Variable and Feature Selection," *Journal of Machine Learning Research*, vol. 3, pp. 1157–1182, 2003.

[3] T. Hastie, R. Tibshirani, and J. Friedman, *The Elements of Statistical Learning: Data Mining, Inference, and Prediction* (2nd ed.), Springer, 2009.

[4] A. Kamilaris, F. X. Prenafeta-Boldú, and A. Likas, "A Review on the Practice of Big Data Analysis in Agriculture," *Computers and Electronics in Agriculture*, vol. 143, pp. 23–37, 2017.

[5] S. P. Mohanty, D. P. Hughes, and M. Salathé, "Using Deep Learning for Image-Based Plant Disease Detection," *Frontiers in Plant Science*, vol. 7, pp. 1–10, 2016.

[6] Y. Ruiz-García and E. Gómez-Plaza, "Elicitors: A Tool for Improving Fruit Phenolic Content," *Agriculture*, vol. 3, no. 1, pp. 33–52, 2013.

[7] A. Kamilaris, A. Kartakoullis, and F. X. Prenafeta-Boldú, "A Review on the Use of Artificial Intelligence in Crop Disease Detection," *Precision Agriculture*, vol. 19, no. 5, pp. 881–905, 2018.

[8] D. Mitra, S. Gupta, D. Srivastava, and S. Sani, "Plant Disease Identification Using Convolution Neural Networks," in D. Srivastava, N. Sharma, D. Sinwar, J. H. Yousif, H. P. Gupta (eds) *Intelligent Internet of Things for Smart Healthcare Systems*, CRC Press, 2023, pp. 203–215.

[9] A. Srinivasan, K. Seetharam, and S. Singh, "Precision Agriculture: An Overview and Advances in Data Analytics for Site-Specific Crop Management," *Computers and Electronics in Agriculture*, vol. 172, 2020.

[10] Y. Zheng, X. Zhang, Q. Cheng, and X. Xu, "A Review on the Applications of Artificial Intelligence in Soil Monitoring and Management," *Journal of Integrative Agriculture*, vol. 20, no. 4, pp. 865–875, 2021.

[11] E. Hassanpour Adeh, F. Abbasi, and H. R. Maier, "A Review of Data-Driven Approaches for Enhancing Water Management and Crop Yield for Irrigation Systems," *Journal of Hydrology*, vol. 568, pp. 285–310, 2019.

[12] A. Mousavi, M. Gharakhani, and A. Mosavi, "Machine Learning in Agriculture: A Review," *Computers and Electronics in Agriculture*, vol. 170, 2020.

[13] A. Fuentes, S. Yoon, and D. S. Park, "A Deep Learning Framework for Groundnut Leaf Disease Identification and Severity Estimation," *Computers and Electronics in Agriculture*, vol. 161, pp. 272–282, 2018.

[14] S. K. Pathan, M. A. Mohammed, and M. H. Anisi, "Internet of Things (IoT) Applications in Precision Agriculture," *Computers and Electronics in Agriculture*, vol. 173, 2020.

[15] S. Shakya, D. Murthy, and M. Kumar, "Crop Yield Prediction Using Machine Learning: A Review," *Computers and Electronics in Agriculture*, vol. 184, 2021.

[16] T. N. N. Lam, V. T. Truong, D. C. Nguyen, D. Q. Le, and T. D. Nguyen, "Applications of Machine Learning in Agriculture: A Survey," *Computers and Electronics in Agriculture*, vol. 175, 2020.

[17] R. Guidotti et al., "A Survey of Methods for Explaining Black Box Models," *ACM Computing Surveys*, vol. 51, no. 5, pp. 1–42, 2018.

[18] I. Lopes and F. Bacao, "A Review on Data Quality Assessment Methods for Machine Learning," *Data & Knowledge Engineering*, vol. 134, 2021.

[19] M. Rabbi, M. A. Rahman, D. R. Kowalski, E. M. Berke, D. B. Flora, and L. C. Brewer, "Protecting Confidentiality of Personal Health Information in Small Health Data Sets," *Journal of Medical Internet Research*, vol. 21, no. 7, 2019.

[20] A. M. Rady and Y. M. Neinaa, "Predicting Yield of Summer Maize Crop using Artificial Neural Networks (ANNs) and Multiple Linear Regression (MLR) Models," *Computers and Electronics in Agriculture*, vol. 186, 2021.

[21] M. Goyal and D. Srivastava, "A Behaviour-Based Authentication to Internet of Things Using Machine Learning," in S. L. Tripathi, D. K. Singh, S. Padmanaban, and P. Raja (eds) *Design and Development of Efficient Energy Systems*, 2022, pp. 245–263.

[22] N. Zhang, R. Wang, and Y. Huang, "Smart Agriculture: An Integrated IoT Framework for Agricultural Decision Making," *IEEE Internet of Things Journal*, vol. 8, no. 3, pp. 1705–1715, 2021.

[23] E. Guzmán, D. Megías, and M. Torres-Torriti, "A Review on Robotics in Agriculture," *Precision Agriculture*, vol. 20, no. 4, pp. 693–719, 2019.

[24] P. Capper, T. Beck, G. Robinson, and J. Arthur, "Agricultural Development and Environmental Policy: A Case for Collaboration in Meeting 21st Century Challenges," *Journal of Environmental Quality*, vol. 47, no. 4, pp. 644–653, 2018.

[25] A. Kamilaris, A. Kartakoullis, and F. X. Prenafeta-Boldú, "A Review on the Use of AI in Agriculture," *Information Processing in Agriculture*, vol. 7, no. 4, pp. 528–550, 2020.

[26] D. G. Tadesse, G. Caramanna, A. Ploeger, A. Kessler, R. Strahan, and K. Naveed, "Artificial Intelligence in Agriculture: A Brief Introduction and Review," *IEEE Access*, vol. 9, pp. 18669–18684, 2021.

[27] S. Fountas et al., "Agricultural Robotics in Support of a Sustainable Rural Development," *Sustainability*, vol. 11, no. 5, 2019.

[28] H. Lütkepohl and M. Schomaker, "Responsible AI for the Food and Agriculture Sector: Overview, Challenges, and Perspectives," *European Journal of Operational Research*, vol. 295, no. 2, pp. 465–475, 2021.

[29] A. Buckwell, J. Mills, J. Ingram, J. Dwyer, and N. Powell, "A Review of Evidence for the Relationship between Farm Business Management Practices and Farm Performance," *Land Use Policy*, vol. 97, 2020.

11

Farming Futures: Leveraging Machine Language for Potato Leaf Disease Forecasting and Yield Optimization

A. Vijayalakshmi, A. Nidin, and Deepthi Das
CHRIST (Deemed to be University), Bangalore, India

11.1 INTRODUCTION

The intersection of agriculture with technology is not a new phenomenon; throughout history, advancements in farming practices have been closely linked with technological innovations. However, the advent of artificial intelligence (AI) and machine learning (ML) has introduced a transformative wave in agriculture that holds the promise of revolutionizing the way we cultivate, manage, and optimize crop production. Traditionally, agriculture has been shaped by a multitude of factors, including weather conditions, soil quality, water availability, and pest and disease outbreaks. These factors contribute to the inherent uncertainties and challenges faced by farmers, often resulting in resource wastage and reduced yields.

Crop yield prediction is a critical facet of modern agriculture, essential for ensuring food security, optimizing resource allocation, and making informed decisions in the face of dynamic environmental conditions. With the world's population steadily growing and climate change introducing greater uncertainty into farming practices, the ability to accurately estimate crop yields has become more crucial than ever. Crop yield prediction leverages a combination of traditional agricultural knowledge, cutting-edge technology, and data-driven analytics to forecast the likely harvest output of various crops, including staples like wheat, rice, corn, and specialty crops such as fruits and vegetables. This predictive capability empowers farmers, policymakers, and the agricultural industry as a whole

DOI: 10.1201/9781003485179-11

to plan effectively, manage resources efficiently, and respond proactively to challenges like pest infestations, disease outbreaks, and adverse weather events. As agriculture increasingly intersects with data science, ML, and remote sensing technologies, the accuracy and sophistication of crop yield prediction models continue to advance, holding the potential to revolutionize global food production and foster sustainable farming practices.

11.2 LITERATURE REVIEW

With the development of big data technologies and high-performance computers, ML has opened up new possibilities for data-intensive science in the multidisciplinary field of agriculture technology [1]. IBM's Watson ML API analyzes weather data, Internet of Things (IoT) sensor data, and other sources to provide farmers with tailored recommendations on planting, harvesting, and irrigation. This technology has been used in various regions to enhance agricultural practices [2]. Google Earth Engine uses satellite imagery and ML to monitor deforestation, track land use changes, and assess crop health worldwide, aiding in agricultural and environmental monitoring [3]. The authors here conducted a review of ML applications in agricultural production systems. The research demonstrates how ML technologies will assist agriculture. The authors in [4] conducted a thorough literature review and analyzed various ML algorithms for predicting plant diseases. Various preprocessing techniques on images are also discussed in this work, which proves how automatic detection can help in increasing the yield of crops. A work carried out by [5] proposed an adaptive learning method, which gave an accuracy of 99.2% on the classification of plant leaf diseases for heterogeneous plant data of rice crops. The authors compared the result with other techniques – namely, back-propagation neural network (BPNN), convolutional neural network (CNN), and support vector machine (SVM) and found that the proposed work gives better accuracy. The analysis was done on 40 years' worth of yield data from 351 German counties by authors in [6] to determine the predominant meteorological and soil hydrological features over four crops. The result of the study proves climatic factors are more important than soil moisture. A literature review conducted on studies that predict crop yield using machine learning algorithms [7] proves that ML algorithms are capable of identifying patterns and extracting information from both structured and unstructured data. The survey also concludes that advanced ML algorithms, like deep-learning methods, have shown they are best at dealing with large amounts of data. The authors in [8] proposed a semi-parametric variant of a deep neural network for data on corn yield from the US Midwest. The results of the model show the negative impacts of climate change on corn yield. In [9], the authors state that there are numerous models for crop classification based on weather, crop disease, growing phase, etc. The authors did a review on various methodologies and the accuracy they provide for agricultural yield prediction using AI techniques.

The comprehensive review of the literature presented above unequivocally highlights the instrumental role that ML algorithms play in the realm of crop yield prediction and their transformative impact on the agriculture sector. Through a myriad of studies and research endeavors, it becomes abundantly clear that ML algorithms serve as powerful tools capable of harnessing the potential of data-driven insights to revolutionize farming practices. These algorithms have demonstrated their capacity to decipher complex patterns, predict crop yields with remarkable accuracy, and provide invaluable support to farmers in making informed decisions related to planting, cultivation, and resource allocation. Furthermore, their ability to detect and manage crop diseases, optimize resource utilization, and adapt to changing environmental conditions underscores their significance in ensuring food security and sustainability in agriculture. In essence, the amalgamation of ML and agriculture offers a promising avenue for the future, where technology is leveraged to enhance productivity, mitigate risks, and propel the agricultural sector toward greater efficiency and resilience.

11.3 OBJECTIVES AND STRUCTURE OF THE CHAPTER

Potatoes are a staple crop globally, providing a significant portion of the world's dietary calories. However, they are highly susceptible to various diseases, such as late blight and early blight, which can devastate yields if left unchecked. Detecting diseases on potato leaves is of paramount importance for achieving increased crop yield and ensuring food security. Timely disease detection allows farmers to take proactive measures, such as targeted pesticide applications or the removal of infected plants, to prevent the spread of diseases throughout the field. By identifying and addressing diseases early, farmers can minimize yield losses, reduce production costs, and maintain the quality of harvested potatoes. Furthermore, disease detection technologies and image analysis enable precise monitoring and management of potato crops, contributing to sustainable agriculture and enhanced global food production.

In this study, the primary focus revolves around the classification of potato leaf diseases. With the ever-growing challenges in agriculture and the imperative need to enhance crop yield and food security, accurately identifying and categorizing diseases affecting potato leaves is of paramount significance. Through the utilization of advanced techniques in image analysis and ML, we aim to develop a robust classification system capable of discerning various potato leaf diseases. This research not only contributes to early disease detection but also facilitates informed decision-making for farmers, allowing for timely interventions and precise management strategies, ultimately bolstering crop health and yield.

11.4 POTATO AS A STAPLE FOOD CROP

Potato, scientifically known as Solanum tuberosum, holds immense significance as a staple food crop worldwide due to its nutritional value, versatility in culinary applications, and capacity to thrive in diverse climates. Understanding the global production of potatoes and their susceptibility to diseases like late blight and early blight is crucial for ensuring food security and addressing the challenges facing potato farming. Potatoes are a rich source of carbohydrates, dietary fiber, vitamins (especially vitamin C and several B vitamins), and essential minerals like potassium and magnesium. They provide a substantial portion of daily caloric intake for millions of people around the world. In this study, we have considered potato leaf images that are affected by diseases like late blight and early blight, and using a deep-learning model, we predicted the occurrence of diseases on leaves.

Late Blight:
Late blight is one of the most destructive potato diseases. It manifests as dark-green to black lesions on leaves, often surrounded by a white,

FIGURE 11.1
Late blight disease on potato leaf.

fluffy mold-like growth on the undersides, as shown in Figure 11.1. Infected plants rapidly deteriorate, and the disease can spread rapidly in cool, humid conditions. Accurate identification of late blight is crucial because it is highly contagious and can lead to significant yield losses if not addressed promptly.

Early Blight:

Early blight primarily affects the lower leaves of potato plants. It presents as circular to irregularly shaped lesions with a dark center and a yellow halo, as shown in Figure 11.2. These lesions can coalesce and lead to extensive defoliation, reducing photosynthesis and tuber quality. Proper classification of early blight helps farmers implement targeted control measures, including fungicides and pruning of affected leaves, to limit the disease's impact on potato yields and tuber quality.

Healthy Leaf:

A healthy potato leaf should have vibrant green coloration without any significant browning, spotting, or lesion, as shown in Figure 11.3. The leaf surface should be smooth and free from abnormal growths or discolorations.

In summary, accurate disease classification in potato crops is a fundamental aspect of sustainable agriculture, enabling farmers to protect their yields, optimize resource utilization, and contribute to global food security. Farmers and researchers continually work to develop disease-resistant potato varieties and implement sustainable farming techniques

FIGURE 11.2
Early blight disease of potato leaf.

FIGURE 11.3
Healthy potato leaf.

to mitigate the impact of these diseases and ensure the reliability of the potato as a global staple crop.

11.5 DATA ACQUISITION AND AUGMENTATION

Potato is one of the most widely cultivated crops in the world, but it is susceptible to various diseases that can significantly reduce crop yield. Early

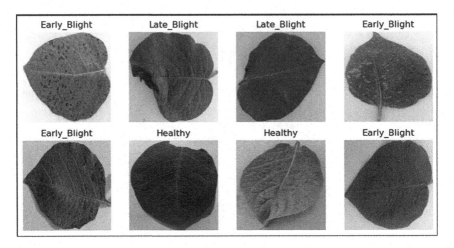

FIGURE 11.4

Sample images from the dataset.

detection and accurate diagnosis of these diseases are crucial for effective disease management. In this chapter, we have considered a case of potato leaf disease that is very common in agriculture. We aim to classify potato leaves using deep-learning techniques for disease detection. We have used the potato leaves images dataset and applied CNNs for classification. A sample of images from the dataset of each category is represented in Figure 11.4.

Various augmentation processes are carried out for the images. Initially, the images are rescaled to a range of [0, 1]. Further, the images are rotated in the range of 30 degrees. Shearing and zoom range are performed by 20%. Width and height wise shift is performed by 5%. The space generated by all these transformations is filled using the nearest neighbor method.

11.6 ML MODELS

Crop yield prediction is a critical component of modern agriculture, offering the potential to enhance food security, optimize resource allocation, and mitigate the impact of climate variability. Deep learning, a subset of ML, has emerged as a powerful tool in this domain, capable of transforming the way we forecast crop yields. Unlike traditional statistical models, deep-learning models can automatically learn and extract complex patterns from vast and diverse agricultural datasets, encompassing factors like weather, soil, and historical yield records. Table 11.1 lists various ML models and their properties [10–13]. By leveraging deep neural networks with multiple layers, these models can capture intricate relationships and spatial-temporal dependencies within the data, leading to more accurate and reliable predictions. In this era of increasing demand for food production and sustainability, deep-learning models for crop yield prediction represent a promising avenue for revolutionizing agricultural practices and ensuring global food security.

TABLE 11.1

List of ML Models in Crop Yield Prediction

Algorithm	Advantages	Disadvantages
Linear Regression	Simple and interpretable, works well with linear relationships	Limited to linear data, sensitive to outliers and multicollinearity
Logistic Regression	Suitable for binary classification Provides probability estimates	Assumes linear decision boundaries May not handle complex relationships well
Decision Trees	Highly interpretable Handles both numerical and categorical data	Prone to overfitting without pruning Sensitive to small data variations
Random Forest	Reduces overfitting through ensemble learning Handles high-dimensional data	May be computationally expensive Less interpretable than a single decision tree
Support Vector Machines	Effective in high-dimensional spaces Can handle nonlinear data with kernel trick	Sensitive to the choice of kernel Can be computationally expensive for large datasets
K-Nearest Neighbors	Simple and easy to implement Nonparametric (no assumptions about data distribution)	Computationally expensive for large datasets Sensitive to the choice of k
Naive Bayes	Efficient and suitable for text classification Works well with high-dimensional data	Assumes independence between features (naive assumption) May not capture complex dependencies
Neural Networks	High capacity for learning complex patterns Suitable for deep-learning tasks	To train neural network, it needs humongous data This can be computationally expensive
K-Means Clustering	Simple and efficient for unsupervised clustering Scales well to large datasets	Requires the specification of the number of clusters (k) Sensitive to initial cluster centroids

11.6.1 Convolutional Neural Network

A CNN, also known as ConvNet, is a deep-learning architecture specifically designed for tasks involving image analysis, recognition, and processing. CNNs have revolutionized computer vision and have been instrumental in numerous applications, from image classification to object detection. CNNs stand out as a dominant architecture for various reasons. Their ability to autonomously extract hierarchical features from data ensures they excel at recognizing patterns and objects within images [14]. CNNs efficiently share parameters across the entire image, making them computationally effective, especially with large datasets. Their translation-invariant property, along with the inclusion of pooling layers, aids in recognizing objects regardless of their position within an image. In combination with data augmentation and scalability, CNNs have solidified their reputation as superior architectures in image classification [15, 16]. Figure 11.5 depicts the basic architecture of CNN.

The various layers in CNN are explained here.

- Input Layer: The input layer of a CNN receives the raw pixel values of an image as its input. Typical input images are 2D arrays with three color channels (Red, Green, and Blue).
- Convolutional Layers: Convolutional layers are the core building blocks of a CNN. They consist of multiple filters (also called kernels) that slide over the input image to extract local features.
- Activation Function: A nonlinear activation function, such as ReLU (Rectified Linear Unit), is applied element-wise after each convolution process to add nonlinearity to the model. ReLU converts all negative values to zero, aiding the network's ability to learn intricate, hierarchical features.
- Pooling Layers: The objective of the pooling layer is to reduce the width and height of the feature map while preserving important information. In a typical pooling technique called max pooling, the

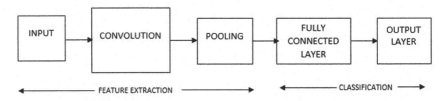

FIGURE 11.5

Architecture of CNN.

maximum value in a particular region of the feature map is kept while the remaining values are discarded.

- Fully Connected Layers: After several convolutional and pooling layers, the network typically ends with one or more fully connected layers. These layers are densely connected, meaning each neuron in one layer is connected to every neuron in the subsequent layer. Fully connected layers learn high-level features and relationships between the features extracted by earlier layers.
- Output Layer: The output layer of a CNN depends on the specific task at hand. For image classification, this typically consists of one neuron per class, followed by a SoftMax activation function to convert the network's output into class probabilities.
- Dropout: Dropout is a regularization technique often applied to fully connected layers to prevent overfitting. During training, a random subset of neurons is "dropped out," meaning their outputs are set to zero, which helps prevent coadaptation of neurons.
- Loss Function: The choice of the loss function depends on the problem being solved. For classification tasks, cross-entropy loss is commonly used, while tasks like regression may use mean squared error.
- Back Propagation and Optimization: CNNs are trained using backpropagation and optimization algorithms like stochastic gradient descent (SGD) or its variants (e.g., Adam, RMSprop). The network updates its weights and biases during training to minimize the chosen loss function.

11.7 MODEL ANALYSIS

Understanding the summary of our model is crucial because it provides vital insights into the architecture, parameters, and behavior of the model we've developed. By analyzing the model summary, we gain an in-depth understanding of how the data flows through the network, the number of trainable parameters, and the layers involved. This knowledge is essential for troubleshooting and fine-tuning the model's performance, diagnosing issues like overfitting or underfitting, and making informed decisions about optimization techniques. Additionally, model summaries help in communication and collaboration within a team or when sharing research findings, ensuring that others can replicate, evaluate, and build upon our work effectively. Table 11.2 describes the summary of the model.

TABLE 11.2

CNN Model Summary on Potato Leaf Dataset

Layer (Type)	Output Shape	Param #
conv2d (Conv2D)	(None, 254, 254, 32)	896
max_pooling2d (MaxPooling2D)	(None, 127, 127, 32)	0
dropout (Dropout)	(None, 127, 127, 32)	0
conv2d_1 (Conv2D)	(None, 127, 127, 64)	18496
max_pooling2d_1 (MaxPooling2D)	(None, 63, 63, 64)	0
dropout_1 (Dropout)	(None, 63, 63, 64)	0
conv2d_2 (Conv2D)	(None, 63, 63, 64)	36928
max_pooling2d_2 (MaxPooling2D)	(None, 31, 31, 64)	0
conv2d_3 (Conv2D)	(None, 31, 31, 64)	36928
max_pooling2d_3 (MaxPooling2D)	(None, 15, 15, 64)	0
conv2d_4 (Conv2D)	(None, 15, 15, 64)	36928
max_pooling2d_4 (MaxPooling2D)	(None, 7, 7, 64)	0
conv2d_5 (Conv2D)	(None, 7, 7, 64)	36928
max_pooling2d_5 (MaxPooling2D)	(None, 3, 3, 64)	0
flatten (Flatten)	(None, 576)	0
dense (Dense)	(None, 32)	18464
dense_1 (Dense)	(None, 3)	99
Total params: 185667 (725.26 KB)		
Trainable params: 185667 (725.26 KB)		
Nontrainable params: 0 (0.00 Byte)		

Conv2D Layer: The model starts with a Conv2D layer with 32 filters, producing an output shape of (none, 254, 254, 32). This is followed by three more Conv2D layers, each with 64 filters. These layers progressively reduce the spatial dimensions.

Max-Pooling Layers: Max-pooling layers follow each convolutional layer, reducing the spatial dimensions by half. They help in spatial downsampling and reducing computational complexity.

Dropout Layers: Dropout layers are used after the max-pooling layers. They help prevent overfitting by randomly dropping a fraction of the input units during training.

Flatten Layer: After the last convolutional layer, there is a flatten layer, which reshapes the 3D output tensor into a 1D vector. This is necessary to connect the convolutional layers to the dense layers.

Dense Layers: Two dense layers follow the flatten layer. The first dense layer has 32 units, and the second dense layer, serving as the output layer, has 3 units, indicating a multiclass classification task.

Total Parameters: The model has a total of 185,667 parameters, all of which are trainable. The parameter count indicates the capacity of the model to learn and represent complex patterns in the data.

Model Capacity: The model's capacity seems adequate for a moderate-sized dataset with three output classes. The 32 units in the first dense layer provide a reasonable level of complexity.

This CNN architecture is used for a multiclass classification task with three classes. It has a reasonable number of parameters and includes essential components like convolutional layers, max-pooling layers, dropout for regularization, and dense layers for classification.

11.8 RESULTS AND DISCUSSIONS

In this section, we present the results of our study in potato leaf disease classification using CNNs. The outcomes of our experiments not only validate the efficacy of this ML approach but also shed light on its potential to revolutionize the field of agricultural disease management. Through rigorous testing and analysis, we have uncovered valuable insights that underscore the significance of our research in enhancing crop yield prediction and, more broadly, in addressing critical challenges in agriculture.

The dataset under study included three folders each for training, testing, and validation. Each of the folders included potato leaf images of all three categories – namely, early blight, late blight, and healthy. For training, we have considered 300 images of early blight, healthy, and late blight potato leaves. One hundred images for testing and 100 for validation. Table 11.3 shows the loss and accuracy of the model.

The training and validation accuracy and loss graphs, depicted in Figure 11.6, serve as vital visual representations of our model's performance throughout the training process. The training accuracy, 0.9856, represents

TABLE 11.3

Loss and Accuracy of the Model

Training Loss	Training Accuracy	Validation Loss	Validation Accuracy	Test Loss	Test Accuracy
0.0470	0.9856	0.0433	0.9965	0.0757	0.9741

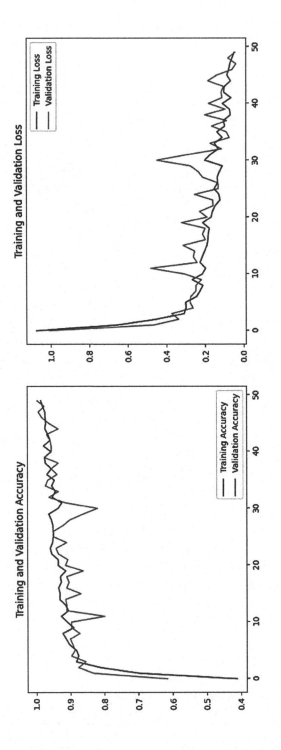

FIGURE 11.6

Training and validation graphs.

the percentage of correctly classified instances within the training dataset as the model learns and adapts.

Simultaneously, the training loss, 0.0470, represents the magnitude of errors made by the model during training. Lower values signify that the model's predictions align more closely with the actual ground truth labels.

On the validation side, we see similar trends. The validation accuracy, 0.9965, shows how well our model generalizes to unseen data, specifically the validation dataset.

In parallel, the validation loss, 0.0433, reflects the model's performance on this separate dataset. As with the training loss, we aim for this value to decrease during training, indicating that the model is not overfitting to the training data but is learning to make more accurate predictions on new, previously unseen data.

The test results provided, including a test loss of 0.0757 and a test accuracy of approximately 97.42%, offer valuable insights into the performance of the CNN model.

1. Test Loss (0.0757):
   The test loss is a crucial metric for assessing the model's predictive accuracy. In this case, the test loss is exceptionally low, indicating that the model's predictions are very close to the actual values in the test dataset. A low test loss suggests that the model has effectively learned the underlying patterns and features in the data, enabling it to make accurate predictions.
2. Test Accuracy (97.42%):
   Test accuracy measures the proportion of correctly classified examples in the test dataset. With an accuracy of approximately 97.42%, the model is performing exceptionally well. An accuracy rate of over 97% suggests that the model is well-suited for the classification task at hand.
3. Model Generalization:
   The impressive test accuracy and low test loss reflect the model's strong generalization capabilities. It is not overfitting the training data, as it can make accurate predictions on new, unseen samples.
4. Model Reliability:
   The high test accuracy and low test loss indicate that the model is reliable and consistent in its predictions.

The test results of the model, with a test loss of 0.0757 and a test accuracy of approximately 97.42%, demonstrate that the model has successfully learned and generalized patterns from the training data. These results suggest that the model is well-suited for the classification task it was designed for.

11.9 CONCLUSION

In conclusion, this study has demonstrated the effectiveness of CNNs in the classification of potato leaf disease, achieving remarkable results with a test loss of 0.0757 and a test accuracy of 0.9741. These findings underscore the potential of ML techniques in agricultural applications, specifically in the early detection and management of plant diseases. By harnessing the power of CNNs, we have paved the way for more accurate and timely identification of potato leaf diseases, enabling farmers to take proactive measures to protect their crops and optimize yields. This research not only contributes to the field of precision agriculture but also holds promise for addressing global food security challenges by improving crop health monitoring and management. As technology continues to evolve, further advancements in ML can lead to even more robust tools for agricultural sustainability and productivity.

REFERENCES

[1] Liakos, Konstantinos G., et al. "Machine learning in agriculture: A review." *Sensors* 18.8 (2018): 2674.

[2] Jain, Ashriti, et al. "Agro-inundation for maximizing crop yield and water efficiency." *SSGM Journal of Science and Engineering* 1.1 (2023): 10–14.

[3] Shelestov, Andrii, et al. "Exploring Google Earth Engine platform for big data processing: Classification of multi-temporal satellite imagery for crop mapping." *Frontiers in Earth Science* 5 (2017): 17.

[4] Ushadevi, G. "A survey on plant disease prediction using machine learning and deep learning techniques." *Inteligencia Artificial* 23.65 (2020): 136–154.

[5] Patel, Bharati, and Aakanksha Sharaff. "Rice crop disease prediction using machine learning technique." *International Journal of Agricultural and Environmental Information Systems (IJAEIS)* 12.4 (2021): 1–15.

[6] Lischeid, Gunnar, et al. "Machine learning in crop yield modelling: A powerful tool, but no surrogate for science." *Agricultural and Forest Meteorology* 312 (2022): 108698.

[7] Bali, Nishu, and Anshu Singla. "Emerging trends in machine learning to predict crop yield and study its influential factors: A survey." *Archives of Computational Methods in Engineering* 29 (2022): 1–18.

[8] Crane-Droesch, Andrew. "Machine learning methods for crop yield prediction and climate change impact assessment in agriculture." *Environmental Research Letters* 13.11 (2018): 114003.

[9] Reddy, D. Jayanarayana, and M. Rudra Kumar. "Crop yield prediction using machine learning algorithm." *2021 5th International Conference on Intelligent Computing and Control Systems (ICICCS)*. IEEE, 2021.

[10] Alzubi, Jafar, Anand Nayyar, and Akshi Kumar. "Machine learning from theory to algorithms: An overview." *Journal of physics: Conference Series*. Vol. 1142. IOP Publishing, 2018.

[11] Kaur, Sunpreet, and Sonika Jindal. "A survey on machine learning algorithms." *International Journal of Innovative Research in Advanced Engineering (IJIRAE)* 3.11 (2016): 2349–2763.

[12] Das, Kajaree, and Rabi Narayan Behera. "A survey on machine learning: Concept, algorithms and applications." *International Journal of Innovative Research in Computer and Communication Engineering* 5.2 (2017): 1301–1309.

[13] Mahesh, Batta. "Machine learning algorithms - A review." *International Journal of Science and Research (IJSR)* 9.1 (2020): 381–386.

[14] Li, Zewen, et al. "A survey of convolutional neural networks: Analysis, applications, and prospects." *IEEE Transactions on Neural Networks and Learning Systems* 33 (2021): 1–21.

[15] O'Shea, Keiron, and Ryan Nash. "An introduction to convolutional neural networks." arXiv preprint arXiv:1511.08458 (2015).

[16] Albawi, Saad, Tareq Abed Mohammed, and Saad Al-Zawi. "Understanding of a convolutional neural network." *2017 International Conference on Engineering and Technology (ICET)*. IEEE, 2017.

12

Classification of Farms for Recommendation of Rice Cultivation Using Naive Bayes and SVM: A Case Study

Qurat-ul-ain, Uzma Hameed, and Hamira Mehraj
Govt. College for Women, Cluster University, Srinagar, India

12.1 INTRODUCTION

The present human population of the world is about 7.6 billion, and it is estimated that the world population will roughly reach 9.8 billion by the end of 2050. Consequently, the food requirements will certainly rise, and contrary to this, the agricultural land is drastically decreasing every passing day due to the miss management in farming, improper planning and unpredictable weather conditions [1], and urbanization. In contemporary times, precision agriculture has gained huge popularity in the agriculture industry with its optimistic results, and it has helped to a great extent in different fields of agriculture like irregular irrigations, plant disease detections, plant counting, crop recommendations, yield estimation, selective harvesting, pest control, and weed detection, as well as many other spheres of agriculture field.

> Precision Agriculture is a management strategy that gathers, processes and analyzes temporal, spatial and individual data and combines it with other information to support management decisions according to estimated variability for improved resource use efficiency, productivity, quality, profitability and sustainability of agricultural production. [2]

DOI: 10.1201/9781003485179-12

In order to gain more insight into data retrieved from precision agriculture, machine-learning (ML) techniques are highly recommended and used in this field. With the implementation of ML tools, we can streamline the process and reduce errors in data, predictions are easier, and interpretation of imagery data becomes quick and easy. "Machine learning is a branch of computer science that broadly aims to enable computers to 'learn' without being directly programmed" [3]. "A machine or an intelligent computer program learns and extracts the knowledge from the data; the extracted knowledge then builds a framework that helps in making predictions or intelligent decisions. Therefore, the ML process is divided into three key parts – i.e., data input, model building, and generalization" [1]. "Generalization is the process for predicting the output for the inputs with which the algorithm has not been trained before" [1]. The ML algorithms can be used to build the classifier model for detecting the quality of soil for proper cultivation of crops. In this chapter, we will understand a similar subject as a case study. To implement a predictive model, let us take a dataset of 2,200 instances and modify it according to the rice crop cultivation and implement two ML predictive models – i.e., Naive Bayes and the support vector machine (SVM) for predicting the rice crop cultivation farm. We will also perform a comparative study of these two models based on various accuracy measures.

12.2 LITERATURE SURVEY

Rice, as we know, is one of the most consumable crops across the globe. Farmers face a lot of difficulties while making a decision in order to be productive and achieve a sustainable crop in the changing climate. "Machine Learning techniques can be used to improve prediction of crop yield under different climatic conditions" [4]. A good number of researchers have applied ML algorithms to detect rice diseases. Jian and Wei have applied SVM for the identification of cucumber leaf disease [5]. Their results have shown that the SVM method based on the radial basis function (RBF) kernel function made the best performance for the classification of diseases of cucumbers as compared to the polynomial and sigmoid kernel functions. Sethy, Barpanda, Rath, and Behera have worked on deep feature-based rice leaf disease detection using CNN and SVM [6].

Gandhi, Petkar, and Armstrong have used neural networks to predict the production yield of rice and to investigate the factors that affect the "rice crop yield for various districts of Maharashtra state in India" [7]. The parameters that were considered by these researchers while carrying out their study were "precipitation, minimum temperature, average temperature, maximum temperature and reference crop evapotranspiration, area, production and yield for the Kharif season (June to November) for the years 1998 to 2002." Singha and et al. have employed the "artificial intelligence (AI) technique for rice and potato crop yield prediction model in the region of Tarakeswar block, Hooghly District, West Bengal." The variables that were predominately used in the study were "climatic factors, static soil parameters, available soil nutrient, agricultural practice parameters, farm mechanization, terrain distribution and socio-economic conditions." They have used "artificial neural network, Support vector machine and Deep Neural network" in their study [8]. The Hasan et al. study has introduced a novel CNN architecture that is comparatively small in size and fairly promising in performance to predict rice leaf disease with moderate accuracy and lower time complexity [9].

Patil and Kumar have proposed a multimodality data fusion framework for rice disease detection.

The Rice-Fusion framework initially extracts the numerical features from agro-meteorological data collected from sensors. Next, it extracts the visual features from the captured rice images. These extracted features are further fused using a concatenation layer followed by a dense layer, which provides single output for diagnosing the rice disease. [10]

Kumar, Singh, Kumar, and P. Singh have proposed a method named the Crop Selection Method (CSM) to solve the crop selection problem and maximize the net yield rate of crops over a season and subsequently achieve maximum economic growth for the country [11]. Son et al. have developed a ML approach for predicting rice crop yields in Taiwan using time-series Moderate Resolution Imaging Spectroradiometer (MODIS) data. Random forests (RF) and SVM were used for building predictive models [12].

12.3 METHODOLOGY

The dataset being used for the study was processed using the WEKA tool. "WEKA (*Waikato Environment for Knowledge Analysis*) [13], is freely available and open source data mining tool available under the GNU General Public License." This tool was developed at the University of Waikato, New Zealand, and was first implemented in its modern form in 1997. In addition to the arff data file format, the WEKA tool also supports .csv data file format, which makes it convenient to export data into different spreadsheet applications. We have used WEKA version 3.8.6 to perform our experiments for the purpose of explanation.

The dataset that has been used as an input to classifiers is available on the Kaggle repository. The name of this dataset is the "crop recommendation" dataset. It is a multivariate dataset with 2,200 instances, and its size is 65 KB. The total number of attributes in this dataset is eight, including the target label attribute. This dataset is classified for different crops, but we have converted the label attribute of the dataset into yes/no according to rice crop recommendations. The selected attributes of the dataset under study are defined in Table 12.1.

12.3.1 Workflow

The input dataset is processed using two ML classifiers – Naive Bayes and SVM. These classifiers are trained and tested for accuracy using the same dataset. Based on the experimental results collected, the best performing

TABLE 12.1

Description of Attributes

S. No.	Attribute Used in Dataset	Type of Attribute	Description
1	N	Numeric	Ratio of nitrogen content in soil
2	P	Numeric	Ratio of phosphorous content in soil
3	K	Numeric	Ratio of potassium content in soil
4	Temperature	Numeric	Temperature in degrees Celsius
5	Humidity	Numeric	Relative humidity in percentage
6	Ph	Numeric	Ph value of the soil
7	Rainfall	Numeric	Rainfall in mm
8	Label	String	Yes/No

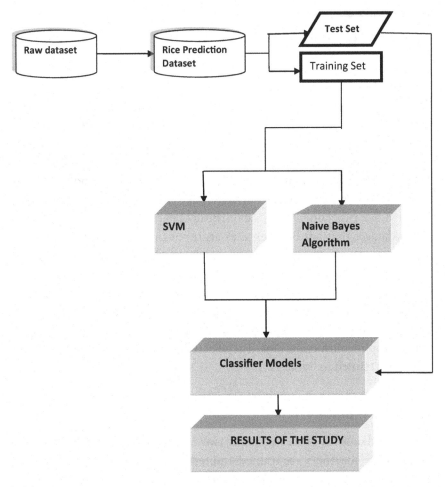

FIGURE 12.1

Flowchart of proposed methodology.

algorithm is determined that will help in the prediction of rice crops to grow on a particular farm based on the parameters depicted in Figure 12.1.

12.3.2 Procedure

The following procedure was followed while building the classifiers and calculating the accuracy of the algorithms using the WEKA tool.

1. The WEKA GUI chooser tool gives the flexibility to choose between Simple CLI, Knowledge Explorer, Experimenter, and Explorer CUI. The experiment was done using the WEKA Explorer interface.

2. Load the input dataset of rice crop recommendation in the WEKA tool.
3. Select the ML algorithm as Naive Bayes or SVM Sequential Minimal Optimization (SMO).
4. Cross-validation was selected as a test option, which is set as ten folds by default.
5. Test the classifier model for the selected algorithm.
6. Perform the comparison of the results obtained from the algorithms.

12.4 CLASSIFIERS EXPLAINED

12.4.1 Naive Bayes

Naive Bayes is a supervised learning classifier that works on the principle of conditional probability. With multiple inputs (x1, x2, x3, x4...xn), the outcome class (Y) using the Naive Bayes equation can be calculated as

$$P(Y = K \mid X_1....X_n) = \frac{\begin{pmatrix} P(X_1 \mid Y = K) * P(X_2 \mid Y = K) * P(X_3 \mid Y = K) \\ *...* P(X_n \mid Y = K) * P(Y = K) \end{pmatrix}}{\left(P(X_1) * P(X_2) * P(X_3)...P(X_n)\right)}. \quad (12.1)$$

In the dataset under study, the Naive Bayes classifier (12.1) is applied to the following attributes:

X = (N, P, K, Temperature, Humidity, Ph, rainfall) and the outcome class Y is classified as Y = (Yes, No).

Different types of evaluation matrices are available in ML (e.g., "Classification Accuracy, Logarithmic Loss, Confusion Matrix, Area under Curve, F1 Score, Mean Absolute Error, and Mean Squared Error"). In this case study, we have considered the Confusion Matrix, Classification Accuracy, and Area under Curve (AUC) evaluation measures to do the comparative analysis of algorithms.

12.4.1.1 Naive Bayes Confusion Matrix

Table 12.2 shows the Confusion Matrix obtained by using the Naive Bayes classifier. Thus, according to this classifier, the values of True Positive = 100, False Negative = 0, False Positive = 43, and True Negative = 2057.

TABLE 12.2

Confusion Matrix Obtained by Using Naive Bayes

Actual Class		Predicted Class	
		Positive	Negative
	Positive	100	0
	Negative	43	2057

12.4.2 Support Vector Machines

SVM is a supervised learning algorithm, which is usually used for classification and regression-based problems. The primary concept of SVMs, initially developed for solving problems related to binary classifications, is the use of hyperplanes to define decision boundaries between data points of different classes. SVMs can handle both simple and linear classification tasks, and can also help in handling more complex – i.e., nonlinear – classification problems. Both separable and nonseparable problems can be well handled by SVMs in linear and nonlinear cases. The idea behind SVMs is to map the original data points from the input space to a high-dimensional, or even infinite-dimensional, feature space such that the classification problem becomes simpler in the feature space. The mapping is done by a suitable choice of a kernel function [14]. In WEKA, we have used a SMO algorithm for training a support vector classifier using a linear kernel on the given dataset in order to build the classification model.

K(x,y) = <x,y> and classifier for class True, False

12.4.2.1 Support Vector Confusion Matrix

Table 12.3 shows the Confusion Matrix obtained by using the support vector classifier. Thus according to this classifier, the values of True Positive = 64, False Negative = 36, False Positive = 3, and True Negative = 2097

TABLE 12.3

Confusion Matrix Obtained by Using SVM

Actual Class		Predicted Class	
		Positive	Negative
	Positive	64	36
	Negative	3	2097

TABLE 12.4

Accuracy Results for Classifiers

Algorithm	Time Taken to Build the Model (sec)	Accuracy Value (%)
Naive Bayes	0.01	98.0455
SVM using WEKA tool	0.11	98.2273

12.5 ACCURACY

Accuracy is another performance evaluation matrix for classification models (Table 12.4). It is calculated as the total number of true predicted instances to the total number of instances presented in the dataset.

$$\text{Accuracy} = (TP + TN)/(TP + TN + FP + FN), \quad (12.2)$$

where in (12.2) TP = True Positive, TN = True Negative, FP = False Positive, FN = False Negative.

Figure 12.2 graphically depicts the accuracy results for the classifiers in which time taken to build the models and accuracy values are shown.

12.6 AREA UNDER CURVE

AUC is another way of measuring the performance of an ML model. The higher the AUC, the better the performance of the model at distinguishing

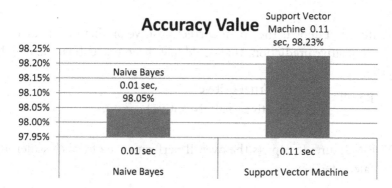

FIGURE 12.2

Graphical representation of accuracy results.

between the positive and negative classes. Following are the classifier accuracy measure values:

  I. True-Positive Rate:
 II. False-Positive Rate:
III. Precision:
IV. Recall:
 V. F-Measure:

  I. True-Positive Rate: It depicts what portion of the positive class gets correctly classified. It is calculated as

$$TruePostive - Rate = \frac{TruePosive}{(TruePositive + FalseNegative)}. \qquad (12.3)$$

 II. False-Positive Rate: It depicts what portion of data is classified as abnormal, which is actually normal.

$$FalsePositive - Rate = FalsePositive / (FalsePositive + TrueNegative) \qquad (12.4)$$

III. Precision: It depicts the number of correct positive predictions made. It is calculated as

$$Precision = \frac{TruePositive}{(TruePositive + FalsePositive)}. \qquad (12.5)$$

IV. Recall: It depicts the no. of correct positive predictions made out of all positive predictions that could have been made. It is calculated as

$$Recall = \frac{TruePositive}{(TruePositive + FalseNegative)}. \qquad (12.6)$$

 V. F-Measure: It depicts the overall performance of the classifier. It is calculated as

$$F - Measure = 2 * (Precision * Recall) / (Precision + Recall). \qquad (12.7)$$

TABLE 12.5

AUC Measures of Algorithms

Algorithm	TP Rate	FP Rate	Precision	Recall	F-Measure
Naive Bayes	0.980	0.001	0.986	0.980	0.982
SVM	0.982	0.344	0.982	0.982	0.981

Table 12.5 represents AUC measures (12.3)–(12.7) of these algorithms obtained after building the classifiers on the dataset in the WEKA tool.

12.7 CONCLUSION

In this chapter, a prediction model based on Naive Bayes and SVM classifiers has been employed to predict whether a particular farm can be recommended for rice cultivation.

The dataset used in carrying out the study for the prediction of farms was obtained from the Kaggle repository, which contains different soil and environmental behavioral attributes.

Naive Bayes and SVM classifiers were built using the WEKA tool. These classifiers thus have been compared based on different accuracy measures to check the performance using the same dataset on two different models. The study led to the conclusion that the TP Ratify Rate and Recall of SVM are higher compared to Naive Bayes. However, the precision and F-measure of Naive Bayes are higher, as compared to SVM. The accuracy value of SVM is 98.2273%, which is slightly better as compared to the Naive Bayes classifier. However, the time taken in building the model in Naive Bayes is 0.01 sec, which is less compared to the time taken using the SVM. Since we should be mainly concerned about the accuracy of the model rather than the time taken in building the model, SVM is the best pick established by the current comparative study.

In the future, we can compare other prediction models like logistic regression, decision support trees, random forest, and neural networks with SVM to establish better prediction models for the current dataset. Moreover, the attribute set can also be enhanced based on various other environmental factors to establish more accurate results so that we may be able to select the right farm for the cultivation of rice crops.

REFERENCES

1 A. Sharma, A. Jain, P. Gupta, and V. Chowdary, "Machine Learning Applications for Precision Agriculture: A Comprehensive Review," *IEEE Access*, vol. 9, pp. 4843–4873, Jan. 2021. doi: 10.1109/access.2020.3048415

2 International Society of Precision Agriculture, "Precision AG Definition|International Society of Precision Agriculture." https://www.ispag.org/about/definition

3 A. L. Samuel, "Some Studies in Machine Learning Using the Game of Checkers," *IBM Journal of Research and Development*, vol. 3, no. 3, pp. 210–229, Jul. 1959. doi: 10.1147/rd.33.0210

4 N. Gandhi, L. J. Armstrong, O. Petkar, and A. K. Tripathy, "Rice Crop Yield Prediction in India Using Support Vector Machines," in *2016 13th International Joint Conference on Computer Science and Software Engineering (JCSSE)*, KhonKaen, Thailand, 2016, pp. 1–5. doi:10.1109/JCSSE.2016.7748856

5 Z. Jian and Z. Wei, "Support Vector Machine for Recognition of Cucumber Leaf Diseases," in *2010 2nd International Conference on Advanced Computer Control*, Shenyang, 2010, pp. 264–266. doi: 10.1109/ICACC.2010.5487242

6 P. K. Sethy, N. K. Barpanda, A. K. Rath, and S. K. Behera, "Deep Feature Based Rice Leaf Disease Identification Using Support Vector Machine," *Computers and Electronics in Agriculture*, vol. 175, p. 105527, Aug. 2020. doi: 10.1016/j.compag.2020.105527

7 N. Gandhi, O. Petkar, and L. Armstrong, "Rice Crop Yield Prediction Using Artificial Neural Networks," in *2016 IEEE Technological Innovations in ICT for Agriculture and Rural Development (TIAR)*, Chennai, India, 2016, pp. 105–110. doi: 10.1109/tiar.2016.7801222

8 C. Singha and K. C. Swain, "Rice and Potato Yield Prediction Using Artificial Intelligence Techniques," in *Studies in Big Data*, 2021, pp. 185–199. doi: 10.1007/978-981-16-6210-2_9

9 M. M. Hasan et al., "Enhancing Rice Crop Management: Disease Classification Using Convolutional Neural Networks and Mobile Application Integration," *Agriculture*, vol. 13, no. 8, p. 1549, Aug. 2023. doi: 10.3390/agriculture13081549

10 R. R. Patil and S. Kumar, "Rice-Fusion: A Multimodality Data Fusion Framework for Rice Disease Diagnosis," *IEEE Access*, vol. 10, pp. 5207–5222, Jan. 2022. doi: 10.1109/access.2022.3140815

11 R. Kumar, M. P. Singh, P. Kumar and J. P. Singh, "Crop Selection Method to Maximize Crop Yield Rate Using Machine Learning Technique," in *2015 International Conference on Smart Technologies and Management for Computing, Communication, Controls, Energy and Materials (ICSTM)*, Avadi, India, 2015, pp. 138–145. doi: 10.1109/ICSTM.2015.7225403

12 N. T. Son et al., "Machine Learning Approaches for Rice Crop Yield Predictions Using Time-Series Satellite Data in Taiwan," *International Journal of Remote Sensing*, vol. 41, no. 20, pp. 7868–7888, Aug. 2020. doi: 10.1080/01431161.2020.1766148

13 "Weka 3 - Data Mining with Open Source Machine Learning Software in Java." https://www.cs.waikato.ac.nz/ml/weka/index.html

14 J. Luts, F. Ojeda, R. Van De Plas, B. De Moor, S. Van Huffel, and J.A. K. Suykens, "A Tutorial on Support Vector Machine-Based Methods for Classification Problems in Chemometrics," *Analytica Chimica Acta*, vol. 665, no. 2, pp. 129–145, Apr. 2010. doi: 10.1016/j.aca.2010.03.030

13

Neural Networks for Crop Disease Detection

Mohammad Ubaidullah Bokhari, Gaurav Yadav,
and Md. Zeyauddin
Aligarh Muslim University, Aligarh, India

13.1 INTRODUCTION

Crop diseases represent a critical and pervasive challenge that has far-reaching implications for food production on a global scale. As the world's population continues to burgeon, the demand for food surges in tandem, necessitating increased agricultural output. However, the insidious threat of crop diseases looms large, consistently undermining efforts to ensure food security. These diseases are a formidable adversary, capable of causing substantial yield losses, reduced crop quality, and economic devastation for farmers and nations alike. They strike indiscriminately, affecting a wide array of crops, from staple grains like rice, wheat, and corn to cash crops such as coffee and cocoa. The consequences of unchecked crop diseases ripple throughout the food supply chain, from reduced crop availability and increased food prices to potential food shortages in vulnerable regions. In a world already grappling with the multifaceted challenges of climate change, resource scarcity, and environmental degradation, the impact of crop diseases exacerbates an already precarious situation. Figure 13.1 illustrates the role of artificial intelligence (AI) in agriculture [1].

In this context, the need for rapid and accurate disease detection in agriculture becomes abundantly clear. Traditional methods of disease identification, reliant on human observation and often requiring laboratory testing, are time-consuming and resource-intensive. By the time a disease outbreak is confirmed, it may have already wrought significant damage.

DOI: 10.1201/9781003485179-13

227

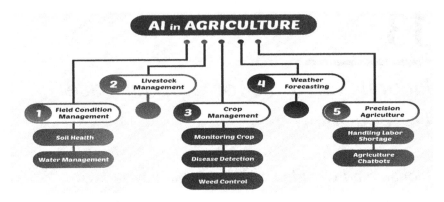

FIGURE 13.1

Role of AI in agriculture.

This lag in detection can hamper the effectiveness of disease management strategies and result in preventable losses. Rapid and accurate disease detection, facilitated by advanced technologies such as AI and remote sensing, offers a transformative solution to this pressing challenge. These technologies empower farmers with the ability to swiftly identify and respond to disease outbreaks, allowing for timely interventions like targeted pesticide application or crop rotation. Furthermore, accurate disease detection enables precision agriculture practices, minimising the unnecessary use of resources and reducing the environmental footprint of agriculture.

13.2 INTRODUCTION OF NEURAL NETWORK

A "neural network" is a kind of computer model that mimics the design and functionality of biological neural networks, such as the human brain. It is composed of neurons, which are interconnected processing units that work together to evaluate data and create judgements. Neural networks have found significant use in a variety of machine learning and AI applications, including image recognition, natural language processing, and others. The same can be understood from the illustration in Figure 13.2. Here are a few kinds of neural networks that are described in depth with illustrations:

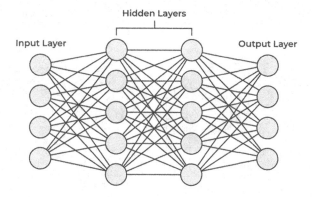

FIGURE 13.2
Working of neural network [1].

1. **Feedforward Neural Network (FNN):** This kind of neural network is the simplest since there are no feedback loops, and information just travels in one way from the input layer to the output layer. As an example, binary classification tasks like email spam detection may be performed with a simple FNN. Email characteristics are received by the input layer; hidden levels analyse them, and the output layer determines whether or not the email is spam.
   - FNNs are the most basic form of neural networks.
   - They are made up of an input layer, one or more hidden layers, and an output layer.
   - Information passes through them in a single path, from the input layer through the hidden layers to the output layer.
   - Commonly used for regression and classification tasks.
2. **CNNs (Convolutional Neural Networks)** are created specifically for image-processing jobs. To automatically extract hierarchical information from input photos, they use convolutional layers. Illustration: A CNN may be used to recognise objects in photos, like as cats or dogs, in image categorisation. Figure 13.3 shows the architecture of CNN.
   - Convolutional layers are used by CNNs to automatically learn spatial hierarchies of features.
   - They analyse grid-like input, such as photos and movies.
   - CNNs have been highly successful in image recognition tasks.
3. **Recurrent Neural Network (RNN):** RNNs are used to process sequential data, and they feature feedback loops that let data flow

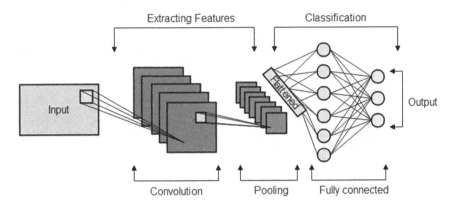

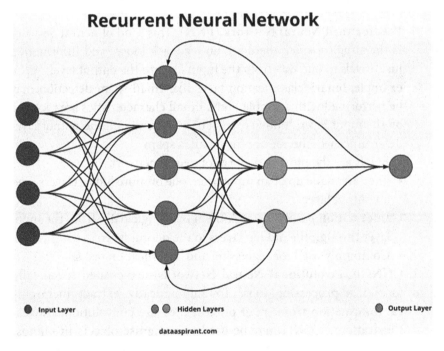

FIGURE 13.3
CNN architecture [1].

FIGURE 13.4
RNN neural network [2].

from one stage in the sequence to the next. Illustration: Because RNNs can preserve context between words in a phrase, they are advantageous for tasks involving natural language processing, including text creation, machine translation, and sentiment analysis. Figure 13.4 shows the architecture of RNN.

- RNNs include recurrent connections, which enable information to be transmitted from one step in the sequence to the next.
- They are specialised for sequential data, such as time series or natural language.
- RNNs are used in tasks like text generation, speech recognition, and machine translation.

4. **Long Short-Term Memory Network (LSTM)**: LSTMs are a specific kind of RNN designed to deal with the vanishing gradient issue. For modelling sequences with long-term dependencies, they work especially well. Illustration: Speech recognition systems may employ LSTMs to translate audio input into text. Figure 13.5 shows the architecture of LSTM.

   - Long sequences are a good fit for LSTMs, a kind of RNN that solves the vanishing gradient issue.
   - They are better suited to handle context because they contain memory cells that can store and retrieve information over lengthy sequences.
   - LSTMs are commonly used in speech recognition, text generation, and language modelling.

5. **Gated Recurrent Unit (GRU)**: GRUs, which are computationally more effective than LSTMs, are intended to capture long-term relationships in sequential data. GRUs can recognise patterns and events in video streams and are utilised in applications like video analysis.

   - GRUs are another type of RNN similar to LSTMs but with a simplified architecture.
   - They have fewer parameters and are often faster to train than LSTMs while performing similarly in many tasks.

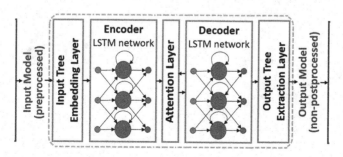

FIGURE 13.5
LSTM neural network [3].

6. **Auto-Encoder**: This kind of neural network is utilised for dimensionality reduction and unsupervised learning. They are made up of a decoder and an encoder. As an example, auto-encoders may be trained to reflect typical data patterns for anomaly identification. Any departure from the established trends can be an abnormality.

   • A decoder reconstructs the input from the encoding, whereas an encoder translates input data to a lower-dimensional representation (encoding).

   • Auto-encoders are employed for unsupervised learning and feature learning.

   • Data compression, denoising, and anomaly detection are used for auto-encoders.

7. **GAN (Generative Adversarial Network)**: A discriminator and a generator are the two neural networks that make up GANs, and they are in competition with one another. The generator creates imaginary data, while the discriminator tries to distinguish between real and fake data. In Figure 13.6, GANs are used in the creation of realistic-appearing pictures, as well as in the transfer of creative styles from one image to another.

   • GANs are composed of two adversarial-trained neural networks: a discriminator and a generator.

   • The generator produces fake data, while the discriminator attempts to tell the difference between actual and fake data.

   • Realistic photos, movies, and other sorts of data may be produced with GANs.

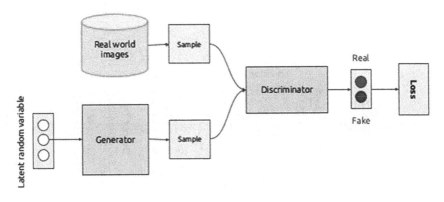

FIGURE 13.6

GAN neural network [4].

The neural network designs that follow are but a few examples. Because of their tremendous adaptability, neural networks may be trained on the right kinds of data and used for a wide range of tasks and domains.

13.3 CONTEMPORARY SCENARIO OF NEURAL NETWORK IN CROP DETECTION

The contemporary scenario of neural networks in crop detection represents a ground-breaking convergence of technology and agriculture, addressing the ever-increasing demand for sustainable food production in the face of mounting global challenges. Previous research in this field has paved the way for transformative developments. For instance, CNNs have emerged as a game-changer in crop disease detection. These deep-learning models have demonstrated remarkable prowess in recognising subtle visual cues and patterns associated with various crop diseases. By analysing images of leaves, fruits, or entire crops, CNNs can swiftly and accurately identify signs of infection or stress, enabling timely intervention and reducing crop losses. Moreover, recurrent neural networks (RNNs) and their variants have proven instrumental in time-series data analysis, essential for monitoring crop health over extended periods. They have been applied in predicting disease outbreaks based on historical weather and disease incidence data, aiding farmers in optimising their disease management strategies. Additionally, GANs have revolutionised the generation of synthetic crop disease images, which, when combined with real data, enhance the robustness of neural network models by exposing them to a broader spectrum of disease manifestations. There are several researchers who have done lots of work in the field of crop disease detection in the field of AI, including machine learning, deep learning, and neural networks. For instance, a study by Shahi et al. [5] intended to analyse current developments in agricultural disease diagnosis with a focus on machine learning and deep-learning methods employing unmanned aerial vehicles (UAV) based remote sensing. First, we discuss how various sensors and image-processing methods are essential for enhancing crop disease detection using UAV data. The second thing we suggest is a taxonomy for compiling and classifying the research already done on agricultural disease identification using UAV photography. The effectiveness of several machine learning and deep-learning techniques for agricultural disease detection is then investigated and summarised. Finally, we highlight the difficulties, prospects, and future research areas of UAV-based remote sensing for agricultural disease identification. In a study, Putra et al. [6] suggested a transfer learning-based approach to dealing with such problems. Our process involves multiple phases. The rice leaf is preprocessed in the first phase. Second, balanced class weighting was used owing to the imbalance in the data. Third, three layers of convolution were added to the transfer learning model to enhance the network performance. Using a bandit-based technique, the parameters of fully linked layers were optimised. The leaf was divided into nine groups in the final stage. We contrast our approach with the most recent state of the art (SOTA) efforts. When compared to the other SOTA, our model has the highest accuracy (98%) at the top.

Apart from this, Jain and Ramesh [7] proposed the work in which CNN and LSTM advantages are combined in the suggested model, which goes by the name hybrid CNN-LSTM. It is a region-specific prediction model that forecasts one-month pest data based on weather and pest data from the previous three months. CNN and LSTM networks are used to compare the performance of the suggested model. This demonstrates how employing hybrid CNN-LSTM has improved performance. In contrast, this study also offers a generalised classification model by fusing the information from every area. Using information from the previous day's pest data and weather, the model forecasts the severity of the illness for the next day. For classifying the severity of an illness, the error correcting output code (ECOC) approach using the support vector machine (SVM) classifier is used. The integration of remote sensing technologies, such as satellites and drones, with neural networks has ushered in an era of precision agriculture. These technologies provide vast amounts of high-resolution data, which neural networks can process to monitor crop conditions, detect anomalies, and assess the impact of environmental factors on crop health. In their study, Fenu and Malloci [8] analyse and categorise research articles over the last ten years that predict the emergence of illness at an early or pre-symptomatic stage (i.e., symptoms that are not apparent to the human eye). We look at the particular techniques and approaches employed, the data and preprocessing methods used, performance indicators, and anticipated outcomes, emphasising the problems discovered. The study's findings show that this practice is still in its infancy and that there are still many obstacles to be solved. In the study by Annabel and Muthulakshmi [9], it is suggested to use a unique method for detecting tomato leaf disease that consists of four steps: picture preprocessing, segmentation, feature extraction, and image classification. The several methods utilised for implementing the suggested technique include thresholding, Gray-Level Co-occurrence Matrix (GLCM), random forest classifier, and Red-Green-Blue (RGB) to grayscale conversion. The findings show that the suggested technique correctly diagnoses the illnesses with a 94.1% accuracy.

Zhong and Zhao [10] proposed a study on apple leaf diseases, which generate significant yearly economic losses and are the primary factor impacting apple output. Studying the diagnosis of apple leaf diseases is, therefore, very important. Three techniques for identifying apple leaf illnesses were developed, including multi-label classification, focus loss function, and regression, all of which were based on the DenseNet-121 deep convolution network. Data modelling and technique assessment in this research were done using the apple leaf image data set, which included 2,462 photos of 6 different apple leaf diseases. In comparison to the conventional multi-classification approach based on cross-entropy loss function, which had an accuracy of 92.29%, the suggested methods' accuracy on the test set was 93.51%, 93.31%, and 93.71%, respectively. This holistic approach to crop detection not only improves disease management but also promotes resource-efficient farming practices. However, contemporary challenges persist, including the need for large and diverse datasets, model interpretability, and the adoption of AI-driven solutions by farmers worldwide. Ahsan et al. [11] give a current review of research that has been done in the last ten years to identify agricultural diseases utilising machine learning, deep learning, image-processing methods, the Internet of Things, and hyperspectral image analysis. Additionally, a comparison study of several methods used to find crop diseases was done. This essay also offers potential solutions to the many difficulties that must be solved. The solutions to these problems are then offered in a number of recommendations. The research also offers a look into the future, which could prove to be a very significant and helpful tool for those studying crop disease identification. Nonetheless, the contemporary scenario of neural networks in crop detection is undeniably promising, offering a potent toolkit for safeguarding global food security, mitigating crop disease risks, and ultimately contributing to a more sustainable and resilient agricultural landscape. As ongoing research continues to refine and expand the capabilities of neural networks in agriculture, the potential for enhanced crop detection and management remains virtually limitless.

13.4 SIGNIFICANCE OF NEURAL NETWORK IN CROP DISEASE DETECTION

The significance of neural networks in crop detection is profound, as they have the potential to revolutionise agriculture by enhancing crop monitoring, disease detection, and yield prediction. Neural networks bring several advantages to this field, as demonstrated by previous research examples:

- **Improved Accuracy**: The identification of agricultural diseases may be made substantially more accurate using neural networks. CNNs were used, for instance, in a study that was published in the journal *Computers and Electronics in Agriculture* to diagnose citrus illnesses from leaf photographs. Their model outperformed conventional approaches with an accuracy rate of over 99%. Farmers that use this degree of precision may identify illnesses early and take prompt action to reduce crop losses.
- **Early Disease Detection**: Neural networks excel at early disease detection, which is crucial for preventing the spread of diseases and minimising crop damage. In another study, a team of researchers utilised RNNs to predict wheat leaf rust disease. Their model could predict the disease's occurrence before it became visually apparent, allowing for targeted interventions and reducing the need for pesticide use.
- **Precision Agriculture**: Neural networks enable precision agriculture by providing fine-grained information about crop health. For instance, research published in *Precision Agriculture* used a combination of CNNs and spectral imaging to detect nitrogen levels in wheat crops. This allowed for precise nitrogen application, reducing fertiliser usage while maintaining crop yield.
- **Automation and Scalability**: Neural networks can automate the monitoring of vast agricultural fields. In a case study conducted by a technology company, drones equipped with deep-learning algorithms were used to identify crop diseases and pests in a large soybean field. This automation not only saved time but also allowed for early interventions, leading to increased yield. Figure 13.7 shows the automation of AI in crop disease detection.
- **Adaptability**: Neural networks can adapt to various crops and diseases. For example, a study in the *Journal of Plant Pathology*

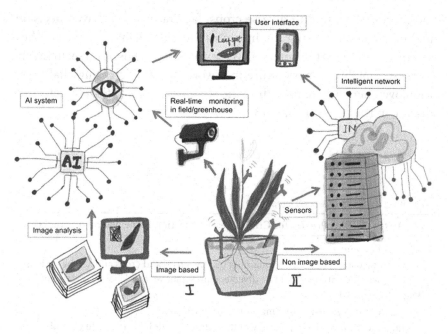

FIGURE 13.7
Significance of neural networks in crop disease detection.

employed deep-learning models to detect multiple diseases across different crop species, including tomatoes, soybeans, and grapes. The adaptability of neural networks makes them valuable tools for diverse agricultural settings.

- **Data-Driven Decision-Making**: Neural networks facilitate data-driven decision-making for farmers. By analysing historical data on crop health and environmental conditions, neural networks can provide insights into optimal planting times, irrigation schedules, and disease management strategies, ultimately improving crop yield and resource utilisation.

- **Reduced Environmental Impact**: Neural network–driven precision agriculture can lead to reduced environmental impact. By optimising resource usage, such as water, pesticides, and fertilisers, these systems contribute to more sustainable farming practices, aligning with the global goal of reducing agriculture's ecological footprint.

Neural networks have demonstrated their significance in crop detection through previous research examples by enhancing accuracy, enabling early disease detection, supporting precision agriculture, automating monitoring, adapting to diverse crops, facilitating data-driven decision-making, and promoting environmentally sustainable practices. These advancements have the potential to improve crop yields, food security, and the sustainability of agriculture in a world facing increasing challenges. Moreover, here we are providing the step-by-step algorithm to detect the disease in the crops in Table 13.1.

TABLE 13.1

Algorithm for Crop Disease Detection

A Step-By-Step Process for Crop Disease Detection Using Neural Networks

1. Data Collection:
- Gather a dataset of images containing healthy crops and crops affected by various diseases. The dataset should be diverse, representative, and labelled accurately.

2. Data Preprocessing:
- Resize and standardise the images to a uniform size.
- Increase the quantity and variety of the dataset by adding changes like rotation, flipping, and brightness tweaks.
- Split the dataset into training, validation, and testing sets (typically, 70%–80% for training, 10%–15% for validation, and 10%–15% for testing).

3. Building the Neural Network:
- Select a neural network design that is appropriate for classifying images. Commonly used for this purpose are CNNs.
- Design the architecture with an input layer, multiple convolutional layers, pooling layers, and fully connected layers.
- Use activation functions like the Rectified Linear Unit (ReLU), and consider using dropout layers to prevent overfitting.
- Select an appropriate loss function, such as categorical cross-entropy, and an optimisation algorithm like Adam.

4. Model Training:
- Feed the training data into the neural network and start training the model.
- Monitor training metrics (e.g., loss and accuracy) on both the training and validation datasets.
- Implement early stopping to prevent overfitting.

5. Hyperparameter Tuning:
- Experiment with different hyperparameters – for example, batch size, learning rate, and network architecture.
- Use techniques like grid search or random search to find the best hyperparameters.

6. Evaluation:
- After training, evaluate the model's performance on the test dataset.
- Calculate metrics such as accuracy, precision, recall, F1-score, and confusion matrix to assess its effectiveness in disease detection.

(Continued)

TABLE 13.1 (CONTINUED)

A Step-By-Step Process for Crop Disease Detection Using Neural Networks

7. Model Fine-Tuning:
- If the initial results are not satisfactory, fine-tune the model by adjusting hyperparameters or modifying the architecture.

8. Interpretability:
- Consider techniques for model interpretability, such as visualisation of feature maps to understand which parts of the images the model is focusing on for disease detection.

9. Deployment:
- Once you are satisfied with the model's performance, deploy it in the field.
- Create a user-friendly interface for farmers or agricultural experts to upload images for disease diagnosis.
- Ensure that the deployed model can handle new and unseen data effectively.

10. Continuous Monitoring and Updates:
- Regularly assess how the deployed model is doing and change it as necessary with fresh information or enhanced algorithms.
- Stay informed about new developments in crop disease detection and neural networks to keep your solution up-to-date.

13.5 ROLE OF NEURAL NETWORK IN CROP DISEASE DETECTION

Neural networks provide a strong and adaptable method for automating the diagnosis and treatment of plant illnesses, which is a major contribution to agricultural disease detection. The following highlights how neural networks help to handle this important agricultural challenge:

- **Accurate Disease Identification:** Neural networks, particularly CNNs, are excellent at jobs requiring visual identification. CNNs can reliably detect the presence of illnesses in agricultural samples by analysing photos of afflicted plants. They gain the ability to see irregularities and delicate visual patterns in fruit, stems, and leaf tissue that may be difficult for the human eye to notice.
- **Early Detection:** Neural networks are capable of enabling early disease detection, allowing farmers to take prompt action to stop the spread of illnesses. Farmers may stop illnesses in their early stages, minimising crop losses, by routinely scanning fields with cameras

or drones fitted with disease detection algorithms based on neural networks.

- **High-Throughput Screening**: Neural networks can quickly analyse enormous amounts of pictures, making it feasible to screen crops in a high-throughput manner. The efficient and economical monitoring of large agricultural fields depends on this skill.
- **Precision Agriculture**: By assisting farmers in more accurate disease control efforts, neural networks allow precision agriculture. Farmers may utilise neural network insights to administer interventions just where they are required, decreasing the usage of chemicals and their negative effects on the environment. This is an alternative to uniformly spraying pesticides or treatments over whole fields.
- **Remote Sensing**: Neural networks may be used with remote sensing technology, such as drones or satellites, to monitor crops over wide geographic regions. In areas where access to agricultural professionals is scarce, this remote monitoring offers useful information for disease identification and control.
- **Continuous Learning**: As disease patterns evolve, neural networks may be taught and retrained using new data. Because of their adaptability, illness detection algorithms may continually learn and become better over time, which raises the accuracy and efficiency of their results.
- **Data Analytics**: Neural networks may be used as part of a wider data analytics pipeline that incorporates information from a variety of sources, including weather, soil quality, and past disease trends. Farmers may get a thorough understanding of disease risks and viable prevention measures from this all-inclusive approach.
- **Scaling Knowledge**: Neural networks offer the ability to scale knowledge and skill in illness diagnosis. Advanced agricultural technology may now be more widely available because of the sharing and deployment of models that have been trained on expert data.
- **Reduced Labour Costs**: Neural networks may eliminate the need for manual labour in crop inspection and disease diagnosis, saving time and money for farmers.

The precise, effective, and scalable solutions offered by neural networks have revolutionised agricultural disease detection. They enable farmers and other agricultural stakeholders to make data-driven choices, reduce crop losses, and promote the growth of healthier and more productive crops, all of which improve global food security. A substantial advancement in efficient and sustainable agriculture may be seen in this technique.

13.6 CONCLUSION

The chapter on the role of neural networks in crop disease detection has, in summary, shed light on a critical future direction for contemporary agriculture. It is impossible to overstate the importance of crop diseases as a worldwide threat to food security, and the use of neural networks for disease detection is a glimmer of hope. With its extraordinary ability for picture recognition, sequence analysis, and data-driven decision-making, neural networks have proved to be very useful tools for both academics and farmers. We have explored the complexities of neural network designs through the lens of this chapter, marvelling at their remarkable capacity to quickly and reliably detect crop illnesses, often in their early stages. The effectiveness of CNNs in image-based detection, the context awareness of RNNs in sequential data processing, and the potential for continuous learning and adaptation in the always-changing fight against crop diseases have all been examined. A new age of precision agriculture has also been ushered in thanks to the convergence of neural networks with cutting-edge technology like remote sensing, drones, and data analytics.

This integration enables the exact implementation of therapies, lowering the environmental effect and increasing resource efficiency, in addition to the early and accurate detection of disorders. The use of neural networks in agricultural disease detection is expected to continue to develop in the future. Technology and knowledge will continue to become more accessible, giving farmers all across the globe access to these effective tools and improving global food security. The chapter has emphasised the enormous potential for scientists, agricultural specialists, and technology to work together in order to push the limits of what is feasible in order to protect our crops and ensure a sustainable and resilient future. They open a new chapter in the history of agriculture, one that is characterised by innovation, empowerment, and optimism. They represent more than simply a chapter in the annals of crop disease detection. This chapter gives us hope for the future because it shows how far neural networks have come and how well-equipped we are to safeguard our crops, feed our people, and create a wealthier and sustainable world.

REFERENCES

1 https://www.edrawsoft.com/article/neural-network-diagram.html

2 https://dataaspirant.com/how-recurrent-neural-network-rnn-works/

3 https://modeling-languages.com/lstm-neural-network-model-transformations/

4 https://medium.com/fnplus/what-is-all-about-generative-adversarial-networks-gans-33faa7b78363

5 Shahi, T. B., Xu, C. Y., Neupane, A., & Guo, W. (2023). Recent Advances in Crop Disease Detection Using UAV and Deep Learning Techniques. *Remote Sensing*, *15*(9), 2450.

6 Putra, O. V., Trisnaningrum, N., Puspitasari, N. S., Wibowo, A. T., & Rachmawaty, E. (2022, August). An Optimized Rice Leaf Disease Classification Using Transfer Learning and Balanced Class Weight Distribution based on Bandit Approach. In *2022 5th International Conference on Information and Communications Technology (ICOIACT)* (pp. 417–422). IEEE.

7 Jain, S., & Ramesh, D. (2021, July). AI Based Hybrid CNN-LSTM Model for Crop Disease Prediction: An ML Advent for Rice Crop. In *2021 12th International Conference on Computing Communication and Networking Technologies (ICCCNT)* (pp. 1–7). IEEE.

8 Fenu, G., & Malloci, F. M. (2021). Forecasting Plant and Crop Disease: An Explorative Study on Current Algorithms. *Big Data and Cognitive Computing*, *5*(1), 2.

9 Annabel, L. S. P., & Muthulakshmi, V. (2019, December). AI-Powered Image-Based Tomato Leaf Disease Detection. In *2019 Third International Conference on I-SMAC (IoT in Social, Mobile, Analytics and Cloud) (I-SMAC)* (pp. 506–511). IEEE.

10 Zhong, Y., & Zhao, M. (2020). Research on Deep Learning in Apple Leaf Disease Recognition. *Computers and Electronics in Agriculture*, *168*, 105146.

11 Ahsan, T., Khan, M., Ahmed, M., Zafar, T., & Javeed, A. (2022). Applications of Artificial Intelligence in Crop Disease Diagnose and Management. *International Journal of Scientific Engineering and Research*, *13*, 1000–1015.

14

Short-Term Weather Forecasting for Precision Agriculture in Jammu and Kashmir: A Deep-Learning Approach

Syed Nisar Hussain Bukhari and Sana Farooq Pandit
National Institute of Electronics and Information Technology (NIELIT),
Srinagar, India

14.1 INTRODUCTION

Agricultural production serves as a primary source of economic development in any country, and weather forecasting (WF) directly or indirectly influences the yearly profitability or loss of agricultural production [1]. WF enables farmers to anticipate and prevent potential damages caused by intense weather phenomena, including reduced yields, misirrigation, erroneous crop selection, and inaccurate harvesting times. Unpredictable weather changes can significantly hinder agricultural progress for an extended period of time. The capacity to anticipate approaching snowfall, rainfall, heatwaves, and floods aids farmers in strategizing and subduing the devastating impacts of such events.

With the advancement of artificial intelligence (AI) in WF, the reliability and accuracy of predictions has significantly improved. Data-driven AI algorithms like neural networks (NN) can be employed to analyse historical weather data and forecast future trends and patterns in weather. These models also improve the accuracy of short-term weather predictions, aiding agriculturists to take informed decisions. Conversely, the performance of statistical techniques such as Autoregressive Moving Average (ARMA) and its variations relies on both the quality of the data and their capability to accurately represent time-series (TS) data exhibiting non-stationary patterns across different seasons [2]. The data-driven modeling systems can be

DOI: 10.1201/9781003485179-14

employed to decrease the computational requirements of Numerical Weather Prediction (NWP) while also accounting for the uncertainty present in observations and system noise [3]. This can be ascribed to the capacity of data-driven models to estimate the intrinsic patterns and dynamics of TS data, even without prior knowledge of the parameters. Specifically, Artificial Neural Networks (ANNs) can be employed for this objective owing to their adaptive characteristics and capacity to learn from previous knowledge [4]. These characteristics make ANNs appealing in diverse fields, particularly when dealing with intricate nonlinear phenomena present in datasets such as weather datasets. Different categories of NN, such as feedforward NN, NN employing back propagation, Recurrent Neural Networks (RNNs), and Long Short-Term Memory (LSTM), adopt a distinct approach to prediction, each offering a unique method for achieving accurate results.

To effectively handle weather data, RNN and LSTM models are considered superior choices due to their ability to efficiently process TS data [5]. Conventional NNs lack the necessary power to capture long-term dependencies. As the number of layers increases, they are prone to encountering the vanishing gradient problem. On the other hand, Recurrent Neural Networks (RNNs) are well-suited for handling TS data with diverse temporal scales, allowing them to store long-term information. For the training of RNN, a TS dataset consisting of daily records of minimum temperature (min temp), maximum temperature (max temp), and rainfall values has been acquired from the Indian Meteorological Department (IMD). This study aims to establish a predictive modeling framework for short-term WF in order to assist farmers in efficient agricultural resource management and plan their farming activities ahead of time. The model will utilize past values within a defined time step as input to forecast min temp, max temp, and rainfall values for the shorter number of days.

14.1.1 Contributions

The primary contributions of this study are summarized as follows:

- To reduce the direct and indirect economic load caused by to reliance of farmers on erroneous predictions of weather
- To proactively reduce potential damages in the agricultural sector resulting from severe weather situations
- To construct a deep-learning (DL) model specifically designed for short-term WF using TS data

- To generate precise weather predictions within a limited timeframe by utilizing localized weather data
- To construct a WF model that demonstrates strong adaptability to evolving trends in weather data and effectively handles extreme weather events while maintaining consistent performance

14.2 RELATED WORK

Various techniques based on both statistical methods and data-driven models have been devised over time to forecast weather conditions by predicting one or more weather-related parameters. Researchers [6] conducted a study to predict rainfall intensity using daily temperature, humidity, wind direction, wind speed, and cloud speed. The prediction models were built using Decision Tree (DT), Support Vector Regression (SVR), and Random Forest (RF), achieving an R^2 score of 0.806, 0.900, and 0.980, respectively. In another study, researchers [7] trained different machine-learning models for rainfall prediction in order to support precision agriculture in Sri Lanka. The models were trained using attributes like rain gauge, relative humidity, average temperature, wind speed, wind direction, solar radiation, and ozone concentration specific to the region of Sri Lanka. It was observed that RF performed best with an accuracy of 89.16%. In their work, researchers [8] combined machine-learning techniques with data mining techniques to develop a rainfall prediction model, and it was observed that Autoregressive Integrated Moving Average (ARIMA) models and NNs provide the best accuracy of 70.5%. In the last decade, RNNs have attracted substantial attention and seen significant progress because of their strong and influential modeling capabilities [9]. In a study, researchers [10] employed daily discharge and rainfall to forecast floods. The research findings suggested the potential use of LSTM, which is a variant of RNN, for the development and administration of real-time flood warning systems. In their experimental study, researchers [1] concluded that the LSTM model outperforms the ARIMA model in predicting minimum and maximum temperature by a root mean squared error (RMSE) score of 1.04. The research carried out by Gumaste and Kadam [11] Placeholder Textemployed the Genetic Algorithm (GA) and Fast Fourier Transform (FFT) for future weather prediction using present data [11]. The models were subsequently deployed through mobile apps to assist farmers in planning their pre- and post-agricultural tasks. To achieve precise weather forecasts for the Dellangu village of Indonesia, researchers developed a short-term weather prediction model utilizing ANNs to help farmers deal with their rice field problems [12]. Another research conducted a comparative analysis between ten base machine-learning regressors and an ensemble model [13]. The analysis suggested that the ensemble model performed better than base models for predicting rainfall, relative humidity, and maximum and minimum temperature. In order to reduce the cost of agricultural frost damage, researchers trained machine-learning models like the convolutional neural network, deep neural network, and RF to predict frost events [14]. The research established that soil temperature is a crucial factor when making predictions beyond 24 hours, whereas for shorter-term forecasts, most of the relevant information is derived from other temperature-related parameters.

14.3 PROPOSED WORK

This study presents an innovative and lightweight WF model specifically designed for accurate short-term predictions of temperature and rainfall in Jammu and Kashmir (J&K). The model is designed to be compatible with stand-alone PCs and can be readily implemented in the geographical region J&K. The proposed system utilizes an advanced DL technique known as RNN that is suitable for sequential data and can enhance the accuracy of the predictions. The following subsection provides an elaborate explanation of the RNN model.

14.3.1 Recurrent Neural Network

A traditional RNN [15] can be thought of as a variation of a feedforward NN that possesses an internal memory component to process the input sequences. RNNs have the ability to learn dependencies from previous terms and predict cumulative measures from variable-length sequences, which include a time component. RNNs are particularly useful for developing prediction models using long-term TS datasets, which can be useful in WF. Unlike traditional deep NN that assume inputs and outputs are unrelated to one another, RNNs generate outputs that are influenced by preceding elements within the sequence. For that, RNNs utilize feedback

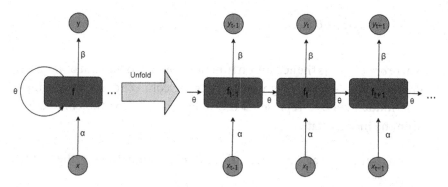

FIGURE 14.1
Working of RNN.

loops to leverage the output from one time step as input for the subsequent step. By maintaining information in a sequential manner, the input of the RNN at each time step incorporates both the data from the current time step in the sequence and the preceding information passed through the feedback loop. The recurrent nature of an RNN stems from its consistent application of the same function to compute the output for each input, with the output being influenced by previous calculations.

Figure 14.1 illustrates a basic RNN architecture in which x_t to x_{t+1} represent the inputs for each sequence, and y_t to y_{t+1} represent the corresponding outputs produced for each sequence. It is evident that all the inputs are interconnected, with each connection denoted by a single RNN cell. The terms f_{t-1} to f_{t+1} refer to hidden states, and θ represents the weight of hidden states; α signifies the weight of the current input state; β represents the weight at the output state. In contrast to traditional feedforward NNs where weights are distinct for each node, RNNs utilize shared parameters across the layers, implying that within each layer of the RNN, the same weight parameter is applied. However, these shared weights are still updated during the backpropagation and gradient descent processes to enable effective reinforcement learning. The functioning of RNN can be explained in the following steps:

For a sequence of data points x, denoted as $x = [x_{t-1}, x_t, x_{t+1}...]$, where x_t represents the input at a specific time step "t" within the sequence:

- The hidden state at time step "t", denoted as f_t, is calculated by combining the current input x_t with the previous hidden state f_{t-1}. This computation is expressed as

$$f_t = \sigma\left(\alpha x_t + \theta f_{t-1}\right),$$

where σ represents the activation function applied to the sum of weighted inputs; α represents the weight matrix for the connections from the input to the hidden state; θ represents the weight matrix for the connections within the hidden state.

- The output y_t at time step "t" can be computed using the hidden state:

$$y_t = \sigma\left(\beta f_t\right)$$

where β is the weight matrix for the hidden-to-output connections.

The activation function is a mathematical operation used to incorporate non-linearity into NN, allowing them to effectively learn intricate data patterns and make nonlinear changes to the input data. In RNNs, the activation function is exclusively applied to the hidden layers since the input layer solely holds the input data and does not undergo any calculations. Additionally, the output of the RNN, which represents the network's prediction or generated output at each time step, can also undergo an activation function to convert the raw output into a preferred range or format based on the specific task or problem being tackled. Without the presence of nonlinear activation functions, a neural network with multiple hidden layers would essentially resemble a large-scale linear regression model and would be unable to effectively learn intricate and complex patterns from real-world data. Commonly used activation functions are Rectified Linear Unit (ReLU), tanh, Sigmoid, and Softmax.

14.3.1.1 Backpropagation through Time (BPTT)

BPTT [16] is an iterative training method utilized specifically for RNNs. It extends the standard backpropagation algorithm, commonly employed for training feedforward NN, and incorporates the temporal aspect of RNNs by allowing gradients to propagate backward across time, considering the sequential nature of the TS data. In RNN, the hidden state of the network at each time step is influenced by both the current input and the output from the previous hidden state. This gives rise to a chain of dependencies that stretches back in time, allowing BPTT to unfold the RNN over time. A simplified breakdown of the steps involved in BPTT is given next:

For a sequence of data points x, denoted as $x = [x_{t-1}, x_t, x_{t+1}...]$, where x_t represents the input at a specific time step "t" within the sequence.

- The hidden state at time step "t", denoted as f_t, is calculated by combining the current input x_t with the previous hidden state f_{t-1}. This computation is expressed as

$$f_t = \sigma\left(\alpha x_t + \theta f_{t-1}\right),$$

Where σ, α, and θ have the usual meanings.

- The loss function, such as mean squared error (MSE) or cross-entropy loss, is applied to calculate the loss at each time step. We can represent the loss at a specific time step "t" as l_t.

  The gradients of the loss with respect to the parameters of the network are calculated from the final time step. Subsequently, these gradients are propagated in a backward direction across previous time steps. The gradient at time step "t," which signifies the impact of the loss on the hidden state at "t" can be represented as:

$$\frac{\partial l_t}{\partial f_t}$$

- The gradients are accumulated at each step as they propagate through time, ensuring that dependencies between different time steps are captured effectively. Accumulated gradients are represented as follows.

$$\frac{\partial l}{\partial f_t}$$

- The parameters of the network are updated by utilizing optimization algorithms such as Adam, SGD and RMSprop. The parameter update for each weight can be computed as

$$\Delta w = -\eta * \Sigma \frac{\partial l}{\partial w},$$

where Δw is the update in weight, and η is the learning rate.

14.4 EXPERIMENTS

Figure 14.2 illustrates the comprehensive process of modeling in detail. The experimentation initiates with the data collection and data preprocessing phase, which is elaborated in the subsequent section. Following that, the process encompasses model configuration, training, comparison, and evaluation. The final step of the experimentation involves assessing and contrasting the performance of the trained models.

14.4.1 Data Collection

This study utilizes real meteorological data obtained from the IMD website [17], specifically from weather stations in Jammu, Srinagar, and Leh. The dataset for each weather station consists of a daily TS comprising 8,401 data points, spanning from January 1, 2010, to December 31, 2022. The dataset includes various weather variables, including rainfall, minimum temperature, and maximum temperature. Rainfall values are measured

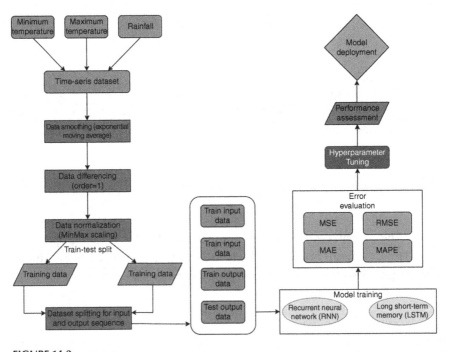

FIGURE 14.2

Illustration of modeling.

in millimeters (mm), while temperature variables are recorded in Celsius. These specific meteorological features are chosen because they provide valuable insights into the current weather conditions at a particular location and time. The selection of these features aims to capture the state of the weather accurately. In this study, these three meteorological features are utilized as input variables for predicting temperature.

14.4.2 Data Preprocessing

The experiment conducted on real data necessitated several preprocessing steps to accurately reflect the performance of the model. In the beginning, smoothing of data has been performed to reduce noise and fluctuations and make underlying patterns and trends more evident. Data has been smoothened using the Exponential Moving Average (EMA) technique. EMA [18] is the weighted average value where weights are decreased gradually such that more importance is given to recent data points as compared to the historical ones, or vice versa. EMA changes at a faster rate and is more sensitive to the data points. Mathematically, the EMA of a data point can be calculated using Equation 14.1.

$$\text{EMA}_t = \begin{cases} x_0 & t = 0 \\ \alpha x_t + (1-\alpha)\text{EMA}_{t-1} & t > 0 \end{cases}. \tag{14.1}$$

where α is a smoothing factor, and its value lies between 0 and 1. It represents the weight applied to the very recent period. For this study, the value of α has been kept at 0.5. To make the TS stationary, differencing of order one has been performed on the weather dataset. Differencing entails computing the differences between consecutive observations by subtracting the previous value from the current value. The purpose of differencing is to eliminate the underlying trend and isolate the specific changes or fluctuations present in the data. For the final stage of preprocessing, data has been normalized using the MinMax scaling technique because the values of different variables in the dataset have different units and lie in multiple ranges. Mathematically, MinMax scaling is represented by Equation 14.2.

$$X_t = \frac{X_t - X_{min}}{X_{max} - X_{min}}, \tag{14.2}$$

where X_t represents the data point at time "t" in TS and X_{max} and X_{min} represent maximum and minimum data points in the sequence, respectively.

Normalized data allows models to generalize well on new and unseen data. Additionally, activation functions in RNN and LSTM neurons work more effectively on normalized data. The dataset has been divided into a training set and a testing set, with proportions of 0.8 and 0.2, respectively. Each weather station's training dataset consists of 6,720 rows containing weather data from January 1, 2000, to May 25, 2018, while the testing dataset contains 1,680 rows spanning from May 26, 2018, to December 31, 2022. Input and output sequences were obtained using the moving window algorithm and a window size of seven. Both input and output sequences contain three features.

14.4.3 Model Configuration and Training

The proposed models have been trained using two distinct configurations: Single Input Single Output (SISO) and Multiple Input Multiple Output (MIMO) [19]. SISO is a forecasting arrangement wherein a single input variable is fed into the network to forecast the future values of the same variable as output. On the other hand, in MIMO, multiple variables are fed into the network in order to predict the same variables as output. Using this configuration, only one model is required per weather station to predict all the future values, while in SISO, three different models are needed for each weather station to forecast three weather parameters. Therefore, three RNN models were trained and evaluated using the MIMO configuration for three distinct weather stations. In addition, a total of nine RNN models were both trained and tested in SISO configuration to forecast three temperature variables across three weather stations. Both configurations have been trained on different sets of hyperparameters of RNN. Hyperparameters are specific values that must be defined prior to model training and are not subject to learning or adjustment during the model training process. The hyperparameters tuned in this study include the number of epochs, batch size, optimization algorithms, and activation functions. The technique chosen for tuning hyperparameters is GridSearchCV, which automates the process of trying various hyperparameter configurations. It thoroughly explores a pre-defined grid of hyperparameter values to identify the combination that produces the highest model performance. For the RNN model, the number of epochs defines how many times the model will be trained on an entire dataset [20–23]. During each epoch, the model goes through one forward pass (input to output) and one backward pass (adjusting weights based on gradients) for all training samples. Using a smaller number of epochs leads to increased error loss, while a larger number of epochs may cause overfitting. The models have been trained for a range of 50–300 epochs. The batch size refers to the number of training samples processed together by the model in one forward and backward pass, allowing the model to update the weights after processing a group of samples. Using a larger batch size during model training comes with higher computational costs, while a smaller batch size introduces noise in the model. Hence, training experiments have been conducted with batch sizes of 16, 32, 64, and 128 to find the optimal configuration. The choice of the optimization algorithm impacts both the speed at which the model reaches its optimal solution and its ability to generalize effectively to unseen data. In order to identify the most suitable optimization algorithm, the models underwent training using Adam, SGD, and RMSprop optimizers. For selecting the optimal activation function, models have been trained on tanh, ReLU, and sigmoid activation functions.

14.4.4 Model Evaluation

The models have been repeatedly trained on training data and evaluated on testing data. For evaluating TS regression models, the commonly used metrics are mean absolute error (MAE), MSE, RMSE, and mean absolute percentage error (MAPE), which are calculated by using Equations 14.3–14.6 [24–26].

$$MAE = \sum_{i=1}^{N} \left| y_t - \acute{y}_t \right| \tag{14.3}$$

$$MSE = \sum_{i=1}^{N} \left(y_t - \acute{y}_t \right)^2 \tag{14.4}$$

$$RMSE = \sqrt{\frac{\sum_{i=1}^{N} \left(y_t - \acute{y}_t \right)^2}{n}} \tag{14.5}$$

$$\text{MAPE} = \sum_{i=1}^{N} \frac{\left| y_t - \dot{y}_i \right|}{y_i} \tag{14.6}$$

where y_t and $\dot{y}_t$ are the actual and predicted values of temperature variables at time step "*t*."

14.5 RESULTS AND DISCUSSION

As mentioned in Section 14.3, the RNN model has been trained using different sets of hyperparameters and tested using various evaluation metrics. For each station, it has been observed that the RNN model performs noticeably better in the SISO configuration than the MIMO configuration. The values of different accuracy metrics for RNN models are listed in Table 14.1 for all weather stations. Tables 14.2–14.4 list the accuracy metrics of the RNN model in SISO configuration for Jammu, Srinagar, and Leh, respectively. These findings indicate that the variables present in weather data are relatively univariate and do not possess strong interdependencies between them. Such relationships can be adequately represented using a single input variable to forecast a single output variable [27–29].

14.6 CONCLUSION AND FUTURE SCOPE

This study highlights the importance of choosing regional data in building a reliable WF model and tackling the WF problem by focusing on

TABLE 14.1

Model Performance in MIMO Configuration for all the Stations

Metrics	Corresponding Value for Jammu Station	Corresponding Value for Srinagar Station	Corresponding Value for Leh Station
MAE	0.0636	0.1048	0.8543
MSE	0.0401	0.0455	0.0411
RMSE	0.1011	0.1555	0.1344
MAPE	0.3990	0.4398	0.3229

TABLE 14.2

Model Performance in SISO Configuration for Jammu Station

Temperature Variables	T_{min}	T_{max}	Rainfall
MAE	0.0751	0.0690	0.0124
MSE	0.0095	0.0095	0.0257
RMSE	0.0977	0.0091	0.0006
MAPE	0.1863	0.1683	0.1229

TABLE 14.3

Model Performance in SISO Configuration for Srinagar Station

Temperature Variables	T_{min}	T_{max}	Rainfall
MAE	0.0658	0.0696	0.0210
MSE	0.0073	0.0926	0.0018
RMSE	0.0854	0.0085	0.0429
MAPE	0.1436	0.1696	0.1368

TABLE 14.4

Model Performance in SISO Configuration for Leh Station

Temperature Variables	T_{min}	T_{max}	Rainfall
MAE	0.0666	0.0085	0.0640
MSE	0.0075	0.0015	0.0078
RMSE	0.0866	0.0393	0.0887
MAPE	0.1547	0.0588	0.1623

achieving precise forecasts without requiring substantial computational resources, in contrast to NWP methods. To resolve this, a lightweight DL-based prediction model has been proposed to ensure accurate short-term WF. The suggested model is capable of operating independently on a single computer unit and can be readily implemented in the geographical area of J&K. In future, more variables like wind speed, humidity, and pressure could be incorporated into the process of model training to provide a more diverse weather forecast. The results also showed that SISO performed better than MIMO configurations; however, that would require three different models to predict three weather parameters as output. In future experiments, MIMO configurations could be further optimized to enhance both performance and efficiency simultaneously.

REFERENCES

[1] Z. Prathyusha, T. Savya, N. Alex, and C. Sobin, "A Method for Weather Forecasting Using Machine Learning," in *2021 5th Conference on Information and Communication Technology, CICT 2021*, Institute of Electrical and Electronics Engineers Inc., 2021. doi: 10.1109/CICT53865.2020.9672403

[2] S. Poornima, and M. Pushpalatha, "Prediction of Rainfall Using Intensified LSTM Based Recurrent Neural Network with Weighted Linear Units," *Atmosphere (Basel)*, vol. 10, no. 11, Nov. 2019. doi: 10.3390/atmos10110668

[3] M. Hayati, and Z. Mohebi, "Application of Artificial Neural Networks for Temperature Forecasting", *International Science Index, Electrical and Computer Engineering*, vol. 1, no. 4, pp. 662–666, 2007.

[4] F. Kratzert, D. Klotz, C. Brenner, K. Schulz, and M. Herrnegger, "Rainfall-Runoff Modelling Using Long Short-Term Memory (LSTM) Networks," *Hydrology and Earth System Sciences*, vol. 22, no. 11, pp. 6005–6022, Nov. 2018. doi: 10.5194/hess-22-6005-2018

[5] T. Kashiwao, K. Nakayama, S. Ando, K. Ikeda, M. Lee, and A. Bahadori, "A Neural Network-Based Local Rainfall Prediction System Using Meteorological Data on the Internet: A Case Study Using Data from the Japan Meteorological Agency," *Applied Soft Computing Journal*, vol. 56, pp. 317–330, Jul. 2017. doi: 10.1016/j.asoc.2017.03.015

[6] V. P. Tharun, R. Prakash, and S. R. Devi, "Prediction of Rainfall Using Data Mining Techniques," in *Proceedings of the International Conference on Inventive Communication and Computational Technologies, ICICCT 2018*, Institute of Electrical and Electronics Engineers Inc., Sep. 2018, pp. 1507–1512. doi: 10.1109/ICICCT.2018.8473177

[7] J. S. A. N. W. Premachandra, and P. P. N. V. Kumara, "A Novel Approach for Weather Prediction for Agriculture in Sri Lanka Using Machine Learning Techniques," in *Proceedings - International Research Conference on Smart Computing and Systems Engineering, SCSE 2021*, Institute of Electrical and Electronics Engineers Inc., Sep. 2021, pp. 182–189. doi: 10.1109/SCSE53661.2021.9568319

[8] U. Shah, S. Garg, N. Sisodiya, N. Dube, and S. Sharma, "Rainfall Prediction: Accuracy Enhancement Using Machine Learning and Forecasting Techniques," in *PDGC 2018 - 2018 5th International Conference on Parallel, Distributed and Grid Computing*, Institute of Electrical and Electronics Engineers Inc., Dec. 2018, pp. 776–782. doi: 10.1109/PDGC.2018.8745763

[9] L. C. Jain and L. R. Medsker, *Recurrent Neural Networks*, 1999. doi: 10.1201/9781420049176

[10] X. H. Le, H. V. Ho, G. Lee, and S. Jung, "Application of Long Short-Term Memory (LSTM) Neural Network for Flood Forecasting," *Water (Switzerland)*, vol. 11, no. 7, 2019. doi: 10.3390/w11071387

[11] S. S. Gumaste and A. J. Kadam, "Future Weather Prediction Using Genetic Algorithm and FFT for Smart Farming." *2016 International Conference on Computing Communication Control and automation (ICCUBEA)*, Pune, India, 2016, pp. 1–6. doi: 10.1109/ICCUBEA.2016.7860028

[12] F. Najib and I. W. Mustika, "Weather Forecasting Using Artificial Neural Network for Rice Farming in Delanggu Village," in *IOP Conference Series: Earth and Environmental Science*, Institute of Physics, 2022. doi: 10.1088/1755-1315/1030/1/012002

[13] C. Nyasulu, A. Diattara, A. Traore, A. Deme, and C. Ba, "Towards Resilient Agriculture to Hostile Climate Change in the Sahel Region: A Case Study of Machine Learning-Based Weather Prediction in Senegal," *Agriculture (Switzerland)*, vol. 12, no. 9, Sep. 2022. doi: 10.3390/agriculture12091473

[14] H. Kerner, M. Alamaniotis, and C. J. Talsma, "Frost Prediction Using Machine Learning and Deep Neural Network Models."

[15] S. Hochreiter, and J. Schmidhuber, "Long Short-Term Memory," *Neural Computation*, vol. 9, no. 8, pp. 1735–1780. 1997. doi: 10.1162/neco.1997.9.8.1735

[16] P. J. Werbos, "Generalization of Backpropagation with Application to a Recurrent Gas Market Model," *Neural Networks*, vol. 1, no. 4, pp. 339–356, 1988. doi: 10.1016/0893-6080(88)90007-X

[17] Home. (n.d.). Accessed: Aug. 07, 2023. https://data.gov.in/catalog/climatology-data-important-cities-0

[18] "Welles Wilder - New Concepts in Technical Trading Systems | PDF | Algorithmic Trading | Financial Economics." Accessed: Aug. 09, 2023. [Online]. Available: https://www.scribd.com/doc/53093880/Welles-Wilder-New-Concepts-in-Technical-Trading-Systems#

[19] F. K. Oduro-Gyimah, and K. O. Boateng, "Forecasting Abilities of MIMO and SISO Neural Networks: A Comparative Study Using Telecommunication Traffic Data," in *Proceedings - 2019 International Conference on Computing, Computational Modelling and Applications, ICCMA 2019*, Institute of Electrical and Electronics Engineers Inc., Mar. 2019, pp. 81–86. doi: 10.1109/ICCMA.2019.00020

[20] S. N. H. Bukhari, J. Webber, and A. Mehbodniya, "Decision Tree Based Ensemble Machine Learning Model for the Prediction of Zika Virus T-Cell Epitopes as Potential Vaccine Candidates," *Scientific Reports*, vol. 12, pp. 7810, 2022. doi: 10.1038/s41598-022-11731-6

[21] G. S. Raghavendra, S. Shyni Carmel Mary, Purnendu Bikash Acharjee, V. L. Varun, Syed Nisar Hussain Bukhari, Chiranjit Dutta, and Issah Abubakari Samori, "An Empirical Investigation in Analysing the Critical Factors of Artificial Intelligence in Influencing the Food Processing Industry: A Multivariate Analysis of Variance (MANOVA) Approach," *Journal of Food Quality*, vol. 2022, Article ID 2197717, 7 pages, 2022. doi: 10.1155/2022/2197717

[22] S. N. H. Bukhari, A. Jain, E. Haq, A. Mehbodniya, and J. Webber, "Ensemble Machine Learning Model to Predict SARS-CoV-2 T-Cell Epitopes as Potential Vaccine Targets," *Diagnostics*, vol. 11, no. 11, 1990. MDPI AG. Retrieved from doi: 10.3390/diagnostics11111990

[23] Sunil L. Bangare, Deepali Virmani, Girija Rani Karetla, Pankaj Chaudhary, Harveen Kaur, Syed Nisar Hussain Bukhari, and Shahajan Miah, "Forecasting the Applied Deep Learning Tools in Enhancing Food Quality for Heart Related Diseases Effectively: A Study Using Structural Equation Model Analysis," *Journal of Food Quality*, vol. 2022, Article ID 6987569, 8 pages, 2022. doi: 10.1155/2022/6987569

[24] C. M. Anoruo, S. N. H. Bukhari, and O. K. Nwofor, "Modeling and Spatial Characterization of Aerosols at Middle East AERONET Stations," *Theoretical and Applied Climatology*, vol. 152, pp. 617–625, 2023. doi: 10.1007/s00704-023-04384-6

[25] F. Masoodi, M. Quasim, S. Bukhari, S. Dixit, and S. Alam, *Applications of Machine Learning and Deep Learning on Biological Data*, CRC Press, 2023.

[26] S. N. H. Bukhari, A. Jain, and E. Haq, "A Novel Ensemble Machine Learning Model for Prediction of Zika Virus T-Cell Epitopes," in Gupta, D., Polkowski, Z., Khanna, A., Bhattacharyya, S., and Castillo, O. (Eds.), *Proceedings of Data Analytics and Management. Lecture Notes on Data Engineering and Communications Technologies*, vol 91. Springer, Singapore, 2022. doi: 10.1007/978-981-16-6285-0_23

[27] S. N. H. Bukhari, F. Masoodi, M. A. Dar, N. I. Wani, A. Sajad, and G. Hussain, "Prediction of Erythemato-Squamous Diseases Using Machine Learning" in *Auerbach Publications eBooks*, pp. 87–96, 2023. doi: 10.1201/9781003328780-6

[28] S. N. H. Bukhari, A. Jain, E. Haq, A. Mehbodniya, & J. Webber, "Machine Learning Techniques for the Prediction of B-Cell and T-Cell Epitopes as Potential Vaccine Targets with a Specific Focus on SARS-CoV-2 Pathogen: A Review," *Pathogens*, vol. 11, no. 2, p. 146, 2022. MDPI AG. doi: 10.3390/pathogens11020146

[29] S. Nisar, H. Bukhari, and M. A. Dar, "Using Random Forest to Predict T-Cell Epitopes of Dengue Virus," *Advances and Applications in Mathematical Sciences*, vol. 20, no. 11, pp. 2543–2547, 2021

15

Deep Reinforcement Learning for Smart Irrigation

Aakansha Khanna
Chandigarh University, Ludhiana, India

Inzimam Ul Hassan
Vivekananda Global University, Jaipur, India

15.1 INTRODUCTION

Data-driven farming is a ground-breaking idea that was born out of the fusion of cutting-edge technology with traditional farming methods in the pursuit of sustainable agriculture and effective resource management. The incorporation of Artificial intelligence (AI) and Machine learning (ML) tools into conventional farming practices is at the core of this change. We explore the many ways that AI and ML are revolutionizing agriculture in the book *Data-Driven Farming: Harnessing the Power of AI and Machine Learning in Agriculture,* and in this chapter, we look at one of its most promising applications: Deep Reinforcement Learning for Smart Irrigation. Water scarcity is a serious problem that affects the entire world, especially the agricultural sector. As the world's population expands, the need for food rises, placing great strain on the planet's already limited water supplies. Finding a balance between the requirement to maintain crop production and ethical water use is a challenge. Conventional irrigation methods, which frequently rely on predetermined schedules or human intuition, can be extremely ineffective, wasting water, raising expenses, and perhaps lowering crop yields. In this situation, Deep Reinforcement Learning (DRL), a branch of ML, offers a revolutionary answer. DRL enables intelligent irrigation systems to make data-driven decisions in real time by enabling systems to learn from and adapt to

DOI: 10.1201/9781003485179-15

changing environmental conditions [1]. This chapter explains how DRL's capacity to optimize irrigation systems greatly aids in the paradigm change in agriculture toward resource conservation and precision farming. We start off by giving a thorough overview of the difficulties posed by conventional irrigation techniques and the potential repercussions of poorly managed water supplies. The fundamentals of DRL are then discussed, showing how it enables machines to discover the best irrigation systems through trial and error while accounting for a variety of environmental characteristics like crop varieties, soil moisture levels, and weather patterns. This chapter will examine real-world case studies and illustrations of successful DRL-based smart irrigation system installation in order to demonstrate the practical advantages these systems offer to farmers, the environment, and the global agricultural landscape [2]. The difficulties and factors to be taken into account when implementing such systems on a broad scale, such as data gathering, sensor technology, and acceptance hurdles, will be covered. We hope to showcase not just the promise of AI and ML in agriculture but also the crucial role they play in addressing some of the most important issues confronting our planet as we set out on this journey into the world of Deep Reinforcement Learning for Smart Irrigation. We open the door to a future that is more sustainable, effective, and productive by utilizing the power of data and technology [3].

DRL has significant potential in the field of agriculture. DRL can be used to maximize crop yield. It can help automate many tasks/subtasks for sustainable farming and lead the way to scalable agriculture solutions [4]. DRL-based AI solutions can be used for crop produce detection. The designed frameworks can be used for automating and controlling greenhouse-based farming processes. For developing a path toward sustainable agriculture, new species of plants are being developed with the help of DRL. These new designs can have far-reaching impacts on agricultural yields and the environment [5]. The deep learning (DL) model has important application value in crop classification, detection, counting, and yield estimation in agriculture. Compared with traditional ML, DL technology improved the performance of hyperspectral image analysis. Here are some key reasons why DRL is important in agriculture:

- **Optimizing Resource Management**: DRL can optimize the allocation of crucial resources in agriculture, such as water, fertilizers, and pesticides. By dynamically adjusting irrigation and nutrient delivery based on real-time data, DRL helps minimize wastage, reduce costs, and maximize resource utilization [6].

- **Precision Agriculture**: DRL enables precision farming by tailoring actions to specific conditions in the field. It considers variables like soil moisture, weather patterns, crop health, and historical data to make informed decisions, ultimately leading to higher crop yields and quality.
- **Water Conservation**: DRL-based irrigation systems make sure that water is applied exactly when and where it is required, lowering water consumption and the environmental impact of farming.
- **Environmental Impact**: DRL contributes to reducing agriculture's environmental footprint. The use of inputs is optimized, and chemical runoff is reduced, which results in more environmentally friendly and sustainable farming methods [5].
- **Crop Yields Can Be Increased**: Using DRL to optimize irrigation, fertilization, and insect management can boost crop yields. DRL assists farmers in producing more with less by ensuring that crops receive the proper supplies at the proper time.
- **Real-Time Decision-Making**: From weather patterns to pest outbreaks, agricultural conditions are always changing. DRL systems enable farmers to react swiftly and successfully to these changes by making decisions in real time.
- **Savings on Labor**: DRL-driven automation can lessen the demand for manual labor in agriculture. Using autonomous drones with DRL algorithms, for instance, can monitor and manage crops while saving labor expenses associated with conventional farming methods.
- **Data-Driven Insights**: DRL generates enormous amounts of data, which may be examined to learn important information about the health of the soil, the detection of diseases, and crop performance. Long-term farming practices and decision-making can be influenced by these insights.
- **Climate Change Adaptation**: The impacts of climate change, such as erratic weather patterns and altering growth seasons, are particularly harmful to agriculture. By adjusting farming methods to changing conditions, DRL can assist farmers in adjusting to these changes [7].

15.1.1 Traditional Irrigation Practices and Their Limitations

Traditional irrigation techniques have been the backbone of agriculture for millennia, supporting civilizations and feeding the world's population.

These traditional techniques, built on the knowledge of previous generations of farmers, have slaked the thirst of crops, converting dry plains into abundant oases and lush green fields. Traditional irrigation techniques have impacted the landscapes and civilizations of our globe in ways that are as fundamental as the dirt that lies beneath our feet. However, they have drawbacks that may impede their effectiveness and sustainability [8]. Fixed-schedule irrigation and intuitive irrigation decision-making are two popular traditional irrigation techniques:

- **Fixed-Schedule Irrigation**: Using fixed-schedule irrigation, farmers water their crops according to preestablished calendars or schedules. Instead of using real-time data, this approach frequently draws on previous conventions and presumptions. Here are a few of its drawbacks:
  - Fixed schedules are inefficient because they don't take into account changes in soil moisture, weather, or crop water requirements. This can result in wasteful water use and possible crop health damage due to over- or underirrigation.
  - Excessive irrigation can lead to excessive water use, which raises operating costs and has negative effects on the environment, such as soil erosion and nutrient leaching.
  - On the other hand, underirrigation can stress crops, reduce yields, and impact overall crop quality.
  - Fixed schedules do not adapt to changing conditions, such as sudden rainfall or unexpected temperature fluctuations. This lack of adaptability can lead to poor irrigation decisions.
- **Intuitive Irrigation Decision-Making**: To decide when and how much to irrigate, farmers use intuitive irrigation decision-making, which is based on their experience, knowledge, and judgment. Even though knowledgeable farmers frequently have a thorough understanding of their crops and terrain, this strategy has drawbacks:
  - Farmers' intuitive judgments are individual and susceptible to change. The precise requirements of the crops or scientific data might not always be in agreement with it.
  - Farmers may occasionally make poor judgments regarding the amount of soil moisture present or fail to account for shifting weather patterns.
  - Relying solely on human judgment can be time-consuming because it necessitates ongoing field inspection and manual adjustments of irrigation systems.

In the realm of traditional irrigation practices, a paradox unfolds: While these methods have nourished crops and civilizations for centuries, they are not without their inherent inefficiencies and consequential challenges. In this section, we will delve into the intricacies of these inefficiencies and the far-reaching consequences they impose on agriculture and the environment [9].

15.1.2 Understanding Deep Reinforcement Learning

In the realm of AI and ML, DRL stands as a beacon of innovation, offering a pathway for machines to not just learn from data but to interact with and adapt to complex environments. Rooted in the fusion of reinforcement learning (RL) principles and deep neural networks, DRL represents a formidable leap forward in the quest to imbue machines with the ability to make sequential decisions, solve intricate tasks, and achieve optimal outcomes in a myriad of domains. DRL is fundamentally a learning-by-doing approach. It makes use of the same idea of trial and error, which has long been the basis of human learning. A digital explorer-like intelligent entity is let loose in a world full of opportunities and obstacles. Its purpose? Take acts, monitor the results, and, via a process of constant improvement, identify techniques that optimize rewards to navigate this terrain. These benefits act as a compass pointing the agent in the direction of its ultimate objective. Some key components of DRL:

Agent: An entity that can perceive/explore the environment and act upon it.

Environment: The world through which the agent moves and where the agent gets its feedback from.

Actions: Steps that the agent takes while moving through the environment.

Rewards: Feedback that the agent gets as a result of its actions. Positive feedback is given for good actions, and negative feedback or penalty is given for bad actions.

Policy: The strategy that the agent employs to determine the next action based on the current state.

Value Function: A prediction of future rewards. It's used to evaluate each state in terms of how good it is for the agent to be in that state.

Q-function or Action-Value Function: Similar to the value function but takes an extra parameter, the current action. Q-function gives the expected future reward for that action taken in the current state [6].

15.1.2.1 Introduction to Deep Learning

A branch of ML called "deep learning" has risen to the top of AI, changing the way that data analysis, pattern detection, and decision-making are done. DL is fundamentally a collection of methods and algorithms drawn from the structure and operation of the human brain, notably neural networks. It excels at processing huge, complicated datasets, allowing computers to identify patterns, forecast outcomes, and carry out previously thought-impossible tasks. DL's defining feature is its utilization of deep neural networks, which are composed of multiple layers of interconnected artificial neurons. These networks have demonstrated remarkable prowess in various domains, including image and speech recognition, natural language processing, autonomous vehicles, and healthcare diagnostics. The hallmark of deep neural networks is their ability to automatically learn intricate features and representations from raw data, eliminating the need for hand-crafted feature engineering [5].

15.1.2.2 Combining Deep Learning and Reinforcement Learning

The marriage of DL and RL emerges as a potent and adaptable fusion in the vast tapestry of AI, where various problems call for versatile answers. The basis for comprehending and displaying complicated patterns in raw data is provided by DL. Its deep neural networks are particularly good at tasks involving perception, including identifying objects in pictures or deciphering text [10]. DL is crucial in the initial phases of decision-making because of its capacity to process sensory input. On the other hand, RL complements DL by focusing on sequential decision-making in dynamic environments. It introduces the concept of an agent interacting with an environment to learn optimal actions through trial and error [11]. RL algorithms aim to maximize a cumulative reward signal over time, making them well-suited for tasks requiring a sequence of actions or strategies. When DL and RL join forces, they create a symbiotic relationship that leverages the strengths of both paradigms:

- **Feature Extraction**: DL can be employed to preprocess raw sensor data (e.g., images, audio, and text) and extract meaningful representations. These representations serve as valuable inputs to RL agents, enhancing their ability to interpret and navigate complex environments.
- **Policy Approximation**: Deep neural networks can be used to approximate the policy function in RL. This allows RL agents to learn

high-dimensional, continuous action spaces efficiently, enabling them to tackle tasks that involve fine-grained control or complex actions.

- **Value Function Approximation**: DL aids in approximating the value functions in RL, which estimate the expected cumulative rewards. Deep Q-Networks (DQNs) and variants like Double DQN leverage deep neural networks to handle high-dimensional state spaces.

- **End-to-End Learning**: The combination of DL and RL enables end-to-end learning, where the entire decision-making pipeline, from perception to action, is learned directly from data. This approach has found success in applications like robotics and autonomous vehicles.

- **Transfer Learning**: DL models pretrained on large datasets can be fine-tuned for RL tasks, allowing agents to benefit from the rich representations learned across various domains. This technique, known as transfer learning, accelerates RL training and enhances its performance [3].

15.1.3 Deep Reinforcement Learning for Smart Irrigation

In the ever-evolving realm of agriculture, the quest for efficient and sustainable irrigation practices has taken on renewed importance. With the global population growing and environmental challenges mounting, the need to optimize the use of water resources in farming has never been greater. Enter DRL, a cutting-edge technology that holds the potential to revolutionize the way we approach smart irrigation. Smart irrigation transcends traditional methods by embracing the power of data-driven decision-making. It leverages DRL, a subset of AI, to transform irrigation into a dynamic and intelligent process. At its essence, DRL for smart irrigation represents a convergence of sophisticated algorithms, real-time data, and agricultural expertise, all working in harmony to achieve the twin objectives of resource efficiency and crop productivity.

- Resource Allocation: At its core, smart irrigation enabled by DRL aims to precisely optimize the distribution of priceless resources like water and nutrients. DRL-enabled systems make intelligent, data-driven decisions by taking into account variables including soil moisture levels, weather patterns, and crop health in real-time.

- Water conservation: DRL-based irrigation systems stand out as a ray of hope in a time when concerns about water scarcity are on the rise. They lessen water waste, reduce environmental impact, and support responsible water conservation in agriculture by providing water precisely where and when it is needed.

- Increased Crop Yields: DRL-driven irrigation surpasses traditional techniques. It orchestrates an ideal growth symphony in addition to keeping crops hydrated. It promotes healthier plants, higher yields, and greater crop quality by ensuring that crops receive the proper materials in the proper quantities.
- Real-Time Adaptation: In the world of agriculture, things can change quickly, from unexpected temperature changes to sporadic rain showers. Farmers can move quickly, accurately, and nimbly through these shifting landscapes thanks to DRL's special ability to make judgments in real time [12].

15.1.3.1 DRL Modification for Agriculture

DRL's application to agriculture represents a successful fusion of innovation into a sector replete with longstanding traditions rather than merely a technology transfer. This process of adaptation is a journey filled with obstacles, chances, and transforming results.

- **Domain Expertise**: Adapting DRL for agriculture necessitates a thorough knowledge of agronomic principles, farming methods, and ecosystems. In order to bridge the gap between theory and practice, the collaboration between domain experts and AI researchers becomes crucial.
- **Data Fusion**: Agriculture creates a vast amount of data, including measurements of soil quality, weather predictions, and crop health evaluations. Utilizing this variety of data and combining it into insights that can be put to use to guide irrigation methods is part of adapting DRL.
- **Sensor Technology**: The deployment of sensor technology is a linchpin in the adaptation process. Accurate and timely data from soil moisture sensors, weather stations, and remote imaging systems empower DRL algorithms to make informed decisions.
- **Scalability**: Agriculture encompasses diverse scales, from small family farms to vast agribusinesses. Adapting DRL involves designing systems that can scale seamlessly to meet the unique needs of each farming operation.
- **Integration with Farming Practices**: The successful adaptation of DRL hinges on its integration into existing farming practices. It

should complement the wisdom of experienced farmers, augmenting their decision-making rather than replacing it [10].

15.1.3.2 Learning Optimal Irrigation Strategies

Learning optimal irrigation strategies through DRL represents the pinnacle of AI's potential in agriculture. It's a journey toward precision, efficiency, and sustainability in crop cultivation.

- **Data-Driven Decision-Making**: DRL systems learn from data-driven experiences, continually optimizing irrigation strategies based on real-time information. This approach replaces intuition with empirical evidence, resulting in superior decision-making.
- **Tailored Resource Allocation**: The quest for optimal irrigation involves the judicious allocation of resources. DRL calculates the precise amount of water, nutrients, and other inputs needed for each section of the field, ensuring resource efficiency and cost savings.
- **Crop-Centric Approach**: Optimal irrigation strategies prioritize the health and growth of crops. DRL ensures that irrigation decisions are aligned with the unique needs of each crop type, growth stage, and environmental condition.
- **Resilience and Adaptability**: The learning process is not static; it adapts to changing conditions. DRL systems are equipped to respond to sudden weather shifts, pest outbreaks, or changes in soil conditions, ensuring crop resilience and yield stability [13].

15.1.3.3 Factors Considered in Decision-Making for Smart Irrigation

In the modern agricultural landscape, the art of irrigation has evolved into a science, guided not only by the wisdom of experienced farmers but also by a wealth of data-driven insights. Smart irrigation, underpinned by advanced technologies and decision-making algorithms, represents a paradigm shift in how we nurture crops, conserve resources, and foster sustainability. At the heart of this agricultural revolution lies the intricate web of factors considered in decision-making for smart irrigation. These factors are the compass points that guide irrigation strategies, transforming what was once a manual, intuition-based practice into a precise, data-driven discipline.

- **Soil Moisture Levels**: Monitoring soil moisture is fundamental to smart irrigation. Sensors embedded in the soil provide real-time data on moisture content, allowing decision-makers to determine if and when irrigation is required.

- **Weather Conditions**: Weather data, including temperature, humidity, wind speed, and precipitation forecasts, play a crucial role in irrigation decisions. Anticipating rain, for example, can lead to irrigation adjustments to avoid overwatering.
- **Crop Type and Growth Stage**: Different crops have varying water requirements at different growth stages. Decision-making considers the specific needs of each crop to optimize irrigation scheduling.
- **Environmental Conditions**: Factors such as evapotranspiration rates (the loss of water from soil and plants), solar radiation, and wind affect water demand and must be factored into irrigation decisions.
- **Topography and Soil Type**: The terrain and soil type of the field influence water distribution. Decision-making accounts for these factors to ensure uniform irrigation across the entire area.
- **Water Source and Availability**: The source of water, be it groundwater, surface water, or irrigation canals, affects availability and cost. Decision-makers consider water source sustainability and associated expenses.
- **Local Regulations**: Agricultural practices are often subject to local regulations governing water use. Compliance with these regulations is essential in decision-making.
- **Cost Analysis**: Economic considerations are vital. Decision-makers evaluate the cost of water, energy for pumping, and irrigation equipment maintenance to optimize resource allocation.
- **Historical Data**: Past irrigation practices and crop performance data provide valuable insights for future decision-making. Historical data helps fine-tune irrigation strategies.
- **Crop Health Monitoring**: Real-time data on crop health, such as disease detection and stress levels, can trigger irrigation adjustments to address plant needs promptly [11].

15.1.3.4 Real-Time Data and Sensing Technologies

In the realm of agriculture, where every moment counts, and every resource is precious, the power of real-time data and sensing technologies has emerged as a beacon of innovation. In the not-so-distant past, farming relied on tradition and experience, but today, it is elevated by an orchestra of sensors, satellites, and interconnected devices that breathe life into the fields. Imagine a world where each acre of land, each plant, and every drop of water speaks its own unique language through data. This symphony of information, orchestrated by a network of sensors and advanced technologies, transcends the boundaries of human perception, offering unprecedented insights into the natural world. Real-time data and sensing technologies are the architects of this transformation. They serve as the eyes and ears of modern agriculture, collecting and relaying a constant stream of information about the environment, crops, and soil conditions. In doing so, they empower farmers and land stewards with the knowledge needed to make informed decisions in a dynamic and ever-changing agricultural landscape. In the pages that follow, we embark on a journey into the heart of real-time data and sensing technologies. We explore the sensors that probe the depths of soil, the satellites that gaze upon our planet from above, and the data analytics that breathe intelligence into raw information. Together, we uncover a world where data is not just a resource but a lifeline – a lifeline that nurtures crops, conserves resources, and charts the course toward a sustainable and prosperous future for agriculture.

- **Soil Moisture Sensors**: These sensors are embedded in the soil to continuously monitor moisture levels at various depths. Real-time data from these sensors guides decisions on when and how much to irrigate.
- **Weather Stations**: Weather stations collect data on temperature, humidity, wind speed, and precipitation. This data is crucial for understanding current weather conditions and predicting future trends.
- **Satellite and Remote Sensing**: Satellite imagery and remote sensing technologies provide a bird's-eye view of crop health, moisture levels, and vegetation indices over large areas. They offer insights into overall field conditions.
- **IoT (Internet of Things) Devices**: IoT devices, including connected weather sensors, soil monitors, and drones, provide a network of real-time data sources that enable decision-makers to make precise adjustments.
- **Data Analytics and AI**: Advanced data analytics and AI techniques process the real-time data streams from sensors to generate actionable insights and optimize irrigation decisions.
- **Mobile Apps and Dashboards**: Decision-makers often access real-time data through mobile applications and web-based dashboards.

These tools provide a user-friendly interface for monitoring and adjusting irrigation remotely.

- **Automation and Control Systems**: Smart irrigation systems are often equipped with automation and control systems that can act on real-time data, adjusting irrigation valves and pumps as needed without manual intervention.
- **Communication Networks**: Reliable communication networks, including wireless and cellular, ensure that real-time data is transmitted from sensors to decision-makers and control systems seamlessly [14].

15.1.4 Applying DRL to Smart Irrigation

Deep Reinforcement Learning (DRL) has shown significant potential in improving irrigation scheduling in various cropping systems. The goal is to discover an intelligent decision rule that processes information available to growers and prescribes sensible irrigation amounts.

For instance, a system called Deep Reinforcement Learning for Irrigation Control (DRLIC) uses a neural network (DRL control agent) to learn an optimal control policy that takes both current soil moisture measurement and future soil moisture loss into account [8].

In another study, an IoT-based irrigation system was proposed, which includes two major components: an IoT wireless network of sensing and actuation nodes and a DRL-based control algorithm. Given the collected soil moisture data and weather data, the DRL-based algorithm finds an optimal irrigation schedule that uses the minimum amount of water to guarantee the soil-water content above the required level before the next irrigation cycle.

These systems have demonstrated their effectiveness in real-world scenarios. For example, the IoT-based system was deployed in a testbed composed of six almond trees. Through a 12-day, in-field experiment, it was found that the proposed system can save up to 7.8% of water over a widely used irrigation scheme.

In summary, DRL can be effectively applied to smart irrigation systems to optimize water usage and improve agricultural efficiency [15].

15.1.4.1 Understanding the Environment: Soil, Weather, Crop Type

Irrigation is highly dependent on three main factors: soil, weather, and crop type.

Soil: The type of soil plays a crucial role in irrigation, as different soils have different water-holding capacities. Soil-based sensors can gather relevant data about volumetric water content, salinity, electrical conductivity, and other crucial parameters. Understanding these parameters can help determine the optimal amount of water required for irrigation.

Weather: Weather conditions significantly affect the irrigation scheduling criteria. Factors such as rainfall, temperature, humidity, wind speed, and solar radiation influence the evapotranspiration rate (the sum of evaporation and plant transpiration). Monitoring these weather conditions can help adjust irrigation schedules to ensure that crops receive the right amount of water at the right time.

Crop Type: Different crops have different water requirements. The crop type determines the irrigation and fertilizer rates. Some crops may be more drought-tolerant than others and may require less frequent watering. Understanding the specific water needs of each crop type can lead to more efficient use of water.

Smart irrigation systems take into account these factors to make better irrigation decisions. They use AI techniques to process data from soil, weather, and plant sensors to optimize water usage [16].

15.1.4.2 Defining the States, Actions, and Rewards

The state in this context represents the current condition of the field that needs to be irrigated. This could include factors such as the current moisture level of the soil, weather conditions (like temperature, humidity, wind speed, and solar radiation), and the type of crop being grown. These factors can be measured using various sensors and weather forecasting data.

The action in this context is the amount of water to be applied to the field at a given time. This could range from no watering at all (in case of sufficient soil moisture or expected rainfall) to heavy watering (in case of dry soil and high evapotranspiration rates). The action is usually determined by a DRL agent based on the current state and the learned policy.

The reward in this context is a measure of how well the action performed in achieving the goal of optimal irrigation. A simple reward function could be the difference between the desired soil moisture level and the actual soil moisture level after irrigation. If the soil moisture level is close to the desired level, a positive reward is given. If it's too dry or too wet, a negative reward is given. The aim of the DRL agent is to maximize the total reward over time, which leads to an optimal irrigation schedule [17].

15.1.4.3 Designing the DRL Agent

Designing a DRL agent for smart irrigation involves several steps.

Problem Formulation: The first step is to formulate the irrigation problem as a Markov Decision Process (MDP). This involves defining the states, actions, and rewards. As discussed earlier, states could include soil moisture levels, weather conditions, and crop types. Actions could be the amount of water to apply, and rewards could be based on how well the irrigation goal is achieved [15].

DRL Model Selection: Choose an appropriate DRL model for the problem. This could be Q-Learning, DQN, Policy Gradients, or any other suitable model. The choice of model depends on the problem characteristics and the available data.

Feature Extraction: Extract relevant features from the state representations. This could involve processing sensor data to obtain useful information about the soil and weather conditions.

Network Architecture Design: Design the neural network architecture for the DRL model. This involves choosing the number of layers, the number of neurons in each layer, the activation functions, etc.

Training: Train the DRL agent using historical data. The agent learns by interacting with the environment (the field), taking actions (applying water), and receiving rewards (feedback on irrigation effectiveness). The goal of training is to learn a policy that maximizes the total reward over time.

Testing and Deployment: After training, test the DRL agent in a simulated or real environment to evaluate its performance. If the performance is satisfactory, deploy the agent for actual use.

Remember that designing a DRL agent requires expertise in RL and DL. It's also an iterative process that involves a lot of trial and error [18].

15.1.5 Successful Applications of DRL in Smart Irrigation

An IoT-based irrigation system was proposed that includes two major components: an IoT wireless network of sensing and actuation nodes and a DRL-based control algorithm. Given the collected soil moisture data and weather data, the DRL-based algorithm finds an optimal irrigation schedule that uses the minimum amount of water to guarantee the soil-water content above the required level before the next irrigation cycle. The system was deployed in a testbed composed of six almond trees. Through a 12-day, in-field experiment, it was found that the proposed system can save up to 7.8% of water over a widely used irrigation scheme.

DRL has considerable potential to improve irrigation scheduling in many cropping systems by applying adaptive amounts of water based on various measurements over time. The goal is to discover an intelligent decision rule that processes information available to growers and pre-scribes sensible irrigation amounts for the time steps considered. The effectiveness of the framework was demonstrated using a case study of irrigated wheat grown in a productive region of Australia where profits were maximized.

DRLIC is a sophisticated irrigation system that uses DRL to optimize its performance. The system employs a neural network known as the DRL control agent, which learns an optimal control policy that considers both the current soil moisture measurement and the future soil moisture loss [19].

15.1.5.1 Current Methods in Smart Irrigation

Smart irrigation systems are a combination of various technologies, such as sensor technology, automatic control technology, and computer technology. They are designed to optimize the efficiency of agricultural water resource utilization.

AI and ML: These techniques are used in both urban and rural agriculture for soil crops to identify systems that are currently being used or can be adapted to urban agriculture.

IoT: IoT allows for more data capture and higher levels of automation. Technologies such as IoT, smartphone tools, and sensors enable farmers to better understand the exact conditions of their fields, including soil temperature, nutrients, water requirements, weather conditions, and more.

Big Data: Modern agricultural operations generate data from a variety of sensors to better understand the environmental factors and the crops, allowing for more accurate and efficient decision-making of farming practices.

Sensor Technology: This includes various types of irrigation systems that can be enhanced with smart irrigation software. The most common are flood, sprinkler, center pivot, drip, and microirrigation systems.

Automatic Control Technology: This is combined with garden irrigation methods, such as sprinkler irrigation and drip irrigation [20].

15.1.5.2 Limitations of Current Methods

While smart irrigation systems have the potential to revolutionize agriculture, they also have certain limitations. The primary disadvantage associated with a smart irrigation system is the expense [4]. These systems can be quite costly, depending on the size of the property. Furthermore, portions of the lawn will have to be dug up to install pipework and attach it to the plumbing system of the home.

Technical Difficulties: Despite the proven benefits of wireless sensor networks in saving water, there are still some technical difficulties, especially in smart irrigation monitoring and control strategies. Even the most efficient smart systems can have their pitfalls. Wind can wreak havoc on sprinklers, directing water in the wrong direction. Underground pests may damage water-delivery systems, resulting in water pooling or broken parts.

Limited Scalability: The research studies on irrigation systems until now are not very efficient, so they cannot be implemented on large-scale systems and have less efficiency due to overburdened sensors for all sensing data [7].

Affordability for Smallholder Farmers: The cost of irrigation technology is often outside of what smallholder farmers can afford as an out-of-pocket expense.

15.1.5.3 Opportunities for Improvement

There are several opportunities for improvement in smart irrigation:

- **Autonomous Pivot Systems**: These systems create new opportunities for smart irrigation.[1] The entire irrigation system is autonomous, and agricultural workers can run the machine from computers, allowing it to stop and start or change water flow [4].
- **Investing in Small-Scale Irrigation**: Small-scale irrigation schemes usually work well, though they must be carefully planned from

the outset so that they are not overwhelmed by complex physical and managerial arrangements.[2] Successful small-scale irrigation is grounded in infrastructure that is efficient, climate-smart, resilient, and financially sustainable [9].

- **Government Initiatives**: The future of the smart irrigation market looks promising, with opportunities in the agriculture and non-agriculture irrigation industries. The major drivers for this market are increasing government initiatives to promote water conservation, growth in smart cities, and increasing the need for efficient irrigation systems [5].
- **Climate Change Adaptation and Improved Food Security**: Improvements in smart irrigation monitoring and management systems may be used to address climate, food, and population issues.
- **Innovative Projects**: There are numerous innovative projects that can be built by students to develop hands-on experience in areas related to/using smart irrigation [12].

15.1.6 Future Directions

DRL has the potential to improve irrigation scheduling in many cropping systems by applying adaptive amounts of water based on various measurements over time. The goal is to discover an intelligent decision rule that processes information available to growers and prescribes sensible irrigation amounts for the time steps considered. The research on this technique remains sparse and impractical due to its technical novelty. To accelerate progress, researchers propose a general framework and actionable procedure that allow researchers to formulate their own optimization problems and implement solution algorithms based on DRL [21–25].

Fresh water is becoming a scarce resource in many parts of the world, and its use in agriculture increasingly needs to be optimized. While there are a number of approaches to irrigation optimization, irrigation scheduling using advanced sensor technologies has considerable potential to apply the right amount of water at the right time based on monitored plant, soil, and atmospheric conditions. The proposed frameworks are general and applicable to a wide range of cropping systems with realistic optimization problems. This means that DRL can be used not only for traditional crops but also for more complex and diverse agricultural systems.

15.1.6.1 Potential Improvements in DRL for Smart Irrigation

DRL has shown significant potential in improving irrigation scheduling in many cropping systems by applying adaptive amounts of water based on various measurements over time [11].

Current research on DRL for irrigation remains sparse and impractical due to its technical novelty. Improvements can be made in handling high-dimensional sensor feedback, which involves processing information available to growers and prescribing sensible irrigation amounts for the time steps considered.

A deep RL-based irrigation scheduling approach can be enhanced to optimize the economic return in irrigation applications. This method involves computing the irrigation quantity for each step while taking evapotranspiration (ET), soil moisture, future precipitation probability, and the current crop growth stage into consideration [16].

The framework for DRL in irrigation is general and applicable to a wide range of cropping systems with realistic optimization problems. However, more research is needed to adapt and fine-tune these models for different types of crops and environmental conditions.

15.1.6.2 Emerging Trends in DRL and Agriculture

DRL is rapidly advancing and has been applied to various agricultural and natural resource management challenges. DRL is being used to solve important problems like crop yield maximization. The goal is to leverage the knowledge available in data and identify ways through which many tasks/subtasks for sustainable farming can be automated, leading to scalable agriculture solutions [18].

DRL is being used for next-best-view planning in agricultural applications. For example, a novel DRL approach has been developed to determine the next best view for the automatic exploration of 3D environments with a robotic arm equipped with an RGB-D camera. This approach improves the information about the regions of interest (ROIs), such as fruits, which are possibly completely or partly occluded by leaves [26–30].

DRL-based solutions are being integrated with IoT wireless networks of sensing and actuation nodes and are being deployed on production-ready cloud infrastructure. This makes the solutions more practical and scalable. These trends indicate that DRL will continue to play a crucial role in advancing agricultural practices, contributing to more efficient farming and sustainable agricultural practices [20].

15.2 CONCLUSION

Within the field of agriculture, the chapter "Deep Reinforcement Learning for Smart Irrigation" provides evidence of how technology can be revolutionary in solving some of the most critical problems facing contemporary farmers. We consider the significant effects that the combination of precision irrigation and DRL has on agriculture as we round out this chapter. Smart irrigation represents a comprehensive strategy for crop cultivation and resource management driven by the intelligence of DRL algorithms. It is evidence of how AI and human expertise work well together, with data-driven choices enhancing the knowledge that has been cultivated over many farming generations. We have examined the variables considered when making decisions about smart irrigation throughout this chapter, highlighting the significance of crop types, weather, soil moisture levels, and sustainability. We have also explored the crucial role that real-time data and sensing technologies – from satellites to soil moisture sensors – play in gathering and utilizing the data required to make these decisions precisely. The combination of DRL with smart irrigation in agriculture has brought increased crop yields, resource efficiency, and sustainability. It has reshaped farming's boundaries, enabling us to preserve the planet's resources while fostering its abundance. It's a tale of creativity, adaptation, and advancement that gets better with every season that goes by. The application of DRL in agriculture continues to have limitless possibilities. In order to improve and broaden the possibilities of smart irrigation, scientists, farmers, and technologists are working together to continue this journey of discovery and invention. The DRL story for smart irrigation is a monument to our dedication to caring for the land and feeding the world, not just one of statistics and algorithms.

REFERENCES

1 Altalak, M., Ammad Uddin, M., Alajmi, A., & Rizg, A. (2022). Smart agriculture applications using deep learning technologies: A survey. *Applied Sciences*, 12(12), 5919.
2 Coulibaly, S., Kamsu-Foguem, B., Kamissoko, D., & Traore, D. (2022). Deep learning for precision agriculture: A bibliometric analysis. *Intelligent Systems with Applications*, 16, 200102.
3 Bhatt, S. (2019, April 19). Reinforcement learning 101. *Medium*. https://towardsdata science.com/reinforcement-learning-101-e24b50e1d292

4 Bu, F., & Wang, X. (2019). A smart agriculture IoT system based on deep reinforcement learning. *Future Generation Computer Systems*, 99, 500–507

5 Zhou, N. (2020, August). Intelligent control of agricultural irrigation based on reinforcement learning. *Journal of Physics: Conference Series*, 1601(5), 052031. IOP Publishing.

6 Yang, Y., Hu, J., Porter, D., Marek, T., Heflin, K., & Kong, H. (2020). Deep reinforcement learning-based irrigation scheduling. *Transactions of the ASABE*, 63(3), 549–556.

7 Alibabaei, K., Gaspar, P. D., Assunção, E., Alirezazadeh, S., & Lima, T. M. (2022). Irrigation optimization with a deep reinforcement learning model: Case study on a site in Portugal. *Agricultural Water Management*, 263, 107480.

8 Debnath, S., Adamala, S., & Palakuru, M. (2020). An overview of Indian traditional irrigation systems for sustainable agricultural practices. *International Journal of Modern Agriculture*, 9, 12–22.

9 Ding, X., & Du, W. (2022, May). Smart irrigation control using deep reinforcement learning. In *2022 21st ACM/IEEE International Conference on Information Processing in Sensor Networks (IPSN)* (pp. 539–540). IEEE.

10 IBM, E. team. (2018). What is deep learning? https://www.ibm.com/topics/deep-learning

11 Fu, Y., Li, C., Yu, F. R., Luan, T. H., & Zhang, Y. (Aug. 2022). Hybrid autonomous driving guidance strategy combining deep reinforcement learning and expert system. *IEEE Transactions on Intelligent Transportation Systems*, 23(8), 11273–11286. doi: 10.1109/TITS.2021.3102432

12 Ganatra, N., & Patel, A. (2021). Deep learning methods and applications for precision agriculture. In *Machine Learning for Predictive Analysis: Proceedings of ICTIS 2020*, pp. 515–527.

13 Kavitha, A. (2021). Deep learning for smart agriculture. *International Journal of Engineering Research & Technology (IJERT) ICRADL*, 09(05), 2021.

14 Attri, I., Awasthi, L. K., Sharma, T. P., & Rathee, P. (2023). A review of deep learning techniques used in agriculture. *Ecological Informatics*, 77, 102217.

15 Sami, M., Khan, S. Q., Khurram, M., Farooq, M. U., Anjum, R., Aziz, S., & Sadak, F. (2022). A deep learning-based sensor modeling for smart irrigation system. *Agronomy*, 12(1), 212.

16 Sun, L., Yang, Y., Hu, J., Porter, D., Marek, T., & Hillyer, C. (2017, December). Reinforcement learning control for water-efficient agricultural irrigation. In *2017 IEEE International Symposium on Parallel and Distributed Processing with Applications and 2017 IEEE International Conference on Ubiquitous Computing and Communications (ISPA/IUCC)*, IEEE, pp. 1334–1341.

17 Gao, H., Zhangzhong, L., Zheng, W., & Chen, G. (2023). How can agricultural water production be promoted? A review on machine learning for irrigation. *Journal of Cleaner Production*, 414, 137687.

18 Çetin, M., & Beyhan, S. (2022). Smart irrigation systems using machine learning and control theory. In *The Digital Agricultural Revolution: Innovations and Challenges in Agriculture through Technology Disruptions*, pp. 57–85.

19 Pallathadka, H., Mustafa, M., Sanchez, D. T., Sajja, G. S., Gour, S., & Naved, M. (2023). Impact of machine learning on management, healthcare and agriculture. *Materials Today: Proceedings*, 80, 2803–2806.

20 Vianny, D. M. M., John, A., Mohan, S. K., Sarlan, A., & Ahmadian, A. (2022). Water optimization technique for precision irrigation system using IoT and machine learning. *Sustainable Energy Technologies and Assessments*, 52, 102307.

21 Bukhari, S. N. H., Webber, J., & Mehbodniya, A. (2022). Decision tree based ensemble machine learning model for the prediction of Zika virus T-cell epitopes as potential vaccine candidates. *Scientific Reports*, 12, 7810. doi: 10.1038/s41598-022-11731-6

22 Raghavendra, G. S., Shyni Carmel Mary, S., Bikash Acharjee, P., Varun, V. L., Bukhari, S. N. H., Dutta, C., & Samori, I. A. (2022). An empirical investigation in analysing the critical factors of artificial intelligence in influencing the food processing industry: A Multivariate Analysis of Variance (MANOVA) approach. *Journal of Food Quality*, 2022, Article ID 2197717, 7 pages. doi: 10.1155/2022/2197717

23 Bukhari, S. N. H., Jain, A., Haq, E., Mehbodniya, A., & Webber, J. (2021). Ensemble machine learning model to predict SARS-CoV-2 T-cell epitopes as potential vaccine targets. *Diagnostics*, 11(11) 1990. MDPI AG. doi: 10.3390/diagnostics11111990

24 Bangare, S. L., Virmani, D., Karetla, G. R., Chaudhary, P., Kaur, H., Bukhari, S. N. H., & Miah, S. (2022). Forecasting the applied deep learning tools in enhancing food quality for heart related diseases effectively: A study using structural equation model analysis. *Journal of Food Quality*, 2022, Article ID 6987569, 8 pages. doi: 10.1155/2022/6987569

25 Anoruo, C. M., Bukhari, S. N. H., & Nwofor, O. K. (2023). Modeling and spatial characterization of aerosols at Middle East AERONET stations. *Theoretical and Applied Climatology*, 152, 617–625. doi: 10.1007/s00704-023-04384-6

26 Masoodi, F., Quasim, M., Bukhari, S., Dixit, S., & Alam, S. (2023). *Applications of Machine Learning and Deep Learning on Biological Data*. CRC Press.

27 Bukhari, S. N. H., Jain, A., & Haq, E. (2022). A novel ensemble machine learning model for prediction of Zika virus T-cell epitopes. In Gupta, D., Polkowski, Z., Khanna, A., Bhattacharyya, S., ans Castillo, O. (Eds.), *Proceedings of Data Analytics and Management. Lecture Notes on Data Engineering and Communications Technologies*, vol 91. Springer, Singapore. doi: 10.1007/978-981-16-6285-0_23

28 Bukhari, S. N. H., Masoodi, F., Dar, M. A., Wani, N. I., Sajad, A. et al. (2023). Prediction of erythemato-squamous diseases using machine learning. In *Applications of Machine Learning and Deep Learning on Biological Data*. CRC Press.

29 Bukhari, S. N. H., Jain, A., Haq, E., Mehbodniya, A., & Webber, J. (2022). Machine learning techniques for the prediction of B-cell and T-cell epitopes as potential vaccine targets with a specific focus on SARS-CoV-2 pathogen: A review. *Pathogens*, 11(2), 146. MDPI AG. doi: 10.3390/pathogens11020146

30 Nisar, S., Bukhari, H., & Dar, M. A. (2021). Using Random Forest to predict T-cell epitopes of Dengue virus. *Advances and Applications in Mathematical Sciences*, 20(11), 2543–2547.

Index

Printed in the United States
by Baker & Taylor Publisher Services